D0728783

INTRODUCTION TO PARTIAL DIFFERENTIAL EQUATIONS

from Fourier Series
to Boundary-value
Problems

ARNE BROMAN

PROFESSOR EMERITUS OF MATHEMATICS
CHALMERS UNIVERSITY OF TECHNOLOGY
GÖTEBORG, SWEDEN

DOVER PUBLICATIONS, INC., NEW YORK

Copyright © 1970 by Addison-Wesley Publishing Company, Inc.
All rights reserved under Pan American and International Copyright
Conventions.

Published in Canada by General Publishing Company, Ltd., 30 Lesmill
Road, Don Mills, Toronto, Ontario.
Published in the United Kingdom by Constable and Company, Ltd.

This Dover edition, first published in 1989, is an unabridged and corrected
republication of the work originally published in 1970 by Addison-Wesley
Publishing Company, Inc., Reading, Mass. Three sections have been
simplified: Lemma 2.4.1, Lemma 4.4.13, and the hint for Exercise 115.
Exercise 132 has been added, and Exercise 421 has been completely changed.
The sections "Conventions" and "Symbols" have been moved from pp. vii–x
to 176–179. The biography on page 184 is also new to this edition.

Manufactured in the United States of America
Dover Publications, Inc., 31 East 2nd Street, Mineola, N.Y. 11501

Library of Congress Cataloging-in-Publication Data

Broman, Arne, 1913–
    Introduction to partial differential equations: from Fourier series to
boundary-value problems / Arne Broman.
        p.   cm.
    "A corrected republication of the work originally published in 1970 by
Addison-Wesley Publishing Company, Inc., Reading, Mass."—T.p. verso.
    Includes bibliographical references.
    ISBN 0-486-66158-X
    1. Differential equations, Partial. I. Title.
QA374.B794   1989
515′.353—dc20   89-16923
                                                          CIP

PREFACE

This text is an outgrowth of a course that has been given every year for some twenty-five years at Chalmers University of Technology, Göteborg, Sweden. The object of the course is to give the students some basic knowledge in Fourier Analysis and in certain of its applications.

The students taking the course have already taken basic courses in calculus, linear algebra, ordinary differential equations, and complex analysis. With such courses as prerequisites, this text is self-contained with respect to Fourier analysis and its applications. In some places where we rely on results from other branches of mathematics that may not have been presented in the basic courses mentioned, references to widely used books have been given.

The definitions and notations used are, as a rule, standard in the mathematical literature of today. In a few places where a definition or a notation is introduced for use in this text only, the word "we" is used in the statement.

The text is divided into eight chapters and subdivided into forty-four sections. At the end of each section exercises are given. These exercises are for the most part divided into two groups: first some routine exercises, and then some more challenging ones, the first of these being marked by an asterisk. The total number of exercises is two hundred and sixty-six.

Formulas of interest have been marked by numbers. Formulas that are not interesting in their own right but to which some reference is given are marked by letters. The first digit in the number of an exercise denotes the chapter in which it occurs.

In order to avoid too many repetitions, some conventions have been introduced, and are collected in an index.

The literature on Fourier analysis and its applications and on the other branches of mathematics that are touched upon in this text is quite extensive. The bibliography indicates a few of the most valuable books in these branches of mathematics.

It is possible for an instructor using this text to omit certain of the sections and starred exercises, and yet to offer a course that should be valuable to the students.

Many people have lectured or given tutorials on this course at Chalmers

iii

during the last two and a half decades. Many ideas in this text emanate from discussions I have had with these people and from lecture notes and collections of exercises that some of them have prepared.

Arne Broman

Göteborg, March 1970

CONTENTS

CHAPTER 1

FOURIER SERIES

1.1 BASIC CONCEPTS

This section (1.1) is devoted to some definitions, notations and conventions concerning intervals, functions and integrals. In what follows we shall adhere to these conventions, etc., unless the context clearly implies some other convention.

The symbol $\in$ means belongs to (is a member of).

1.1.1 Definitions

The set of all real numbers will be denoted by R. Suppose that a and b belong to R and that $a < b$. The sets of all $x \in R$ such that $a < x < b, a \leqslant x \leqslant b, a < x \leqslant b$ or $a \leqslant x < b$ will be denoted by

$$(a, b), \quad [a, b], \quad (a, b], \quad [a, b) \tag{1}$$

respectively. The sets (1) are called *open interval, closed interval* and *half-open intervals* respectively. The terms *interval* and *finite interval* denote any among the sets (1). The number $b - a$ is called the *length* of each of the intervals (1).

In many contexts it is immaterial which among the intervals (1) is considered. In such situations we shall simply write "the interval (a, b)", and in general it is left to the reader to see that our arguments apply equally to any of the intervals (1).

The sets of all $x \in R$ such that $a < x$, $a \leqslant x$, $x < a$, $x \leqslant a$, or x is any real number are denoted by

$$(a, \infty), \quad [a, \infty), \quad (-\infty, a), \quad (-\infty, a], \quad (-\infty, \infty) \tag{2}$$

respectively. The sets (2) are called *infinite intervals*. The second and fourth among them are *half-open infinite intervals*. The remaining three are *open infinite intervals*.

1.1.2 Definitions

A *real-valued function of a real variable* is a set of ordered pairs of real numbers (x, y) such that no two pairs have the same first component. A *function f*, in this chapter, is a real-valued function of a real variable. In this text the set of its first components x, the *domain* of f, is a finite or infinite interval, or such an interval with at most a finite number of x-values missing from each of its finite sub-intervals. The set of its second components y is the *range* of f. The domain and the range of f are denoted by D_f and R_f respectively. A function f is *bounded* if R_f is a subset of some finite interval. A function that is not bounded is called *unbounded*. If $x \in D_f$, the number y in the corresponding pair (x, y) is often denoted by $f(x)$, and the number y or $f(x)$ is called the *value* of the function f at the point x. We shall, however, often let $f(x)$ or $y = f(x)$ denote the function f; it will always be clear from the context which of the two meanings of $f(x)$ is used. If a function consists of all those pairs (x, y) that satisfy a certain equation, we shall often use this equation to define the function; e.g. the phrase "the function $y = \sin x$" denotes the function $\{(x, y); x \in R$ and $y = \sin x\}$. Further, we shall often let expressions such as "the function $y = \sin x$, $0 < x < 2\pi$" define a function; in this example the function $\{(x, y); 0 < x < 2\pi$ and $y = \sin x\}$.

Many authors of mathematical texts define a function as a rule that associates a unique number y with each number x belonging to a certain set. These authors use the term graph (of a function) where we use the term function.

The symbol $\Rightarrow$ means implies (has as a consequence).

Suppose that S is a nonempty set of real numbers. Also suppose that S has an upper bound c, i.e. c is a real number such that $y \in S \Rightarrow y \leqslant c$. Then S has a least upper bound (see [13], Theorem 1.36).* This least upper bound is called the supremum of S, and is denoted by sup S or $\sup_{y \in S} y$. Analogously, if S has a lower bound, then S has a greatest lower bound, called the infimum of S, and denoted by inf S or $\inf_{y \in S} y$.

1.1.3 Definitions

Suppose that f is a bounded function whose domain includes the interval (a, b) with the possible exception of finitely many points (Fig. 1.1). Let D be a subdivision of the interval (a, b), i.e. a strictly increasing finite sequence of numbers x_v, where $v = 0, 1, \ldots, n$, such that $x_0 = a$ and $x_n = b$. (To index finite sequences, Greek letters will be used.) Let the numbers m_v and M_v be defined by the equalities

$$m_v = \inf_{x \in I_v} f(x) \qquad \text{and} \qquad M_v = \sup_{x \in I_v} f(x),$$

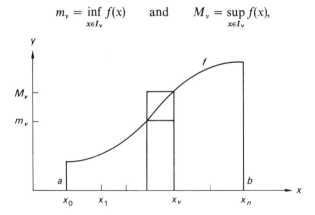

Figure 1.1

---

* Numbers within brackets refer to the bibliography on p. 174.

where I_v denotes the set $\{x; x_{v-1} < x < x_v$ and $x \in D_f\}$. Consider the sums $s(D)$ and $S(D)$ defined by the equations

$$s(D) = \sum_{v=1}^{n} m_v(x_v - x_{v-1}) \qquad \text{and} \qquad S(D) = \sum_{v=1}^{n} M_v(x_v - x_{v-1})$$

(the lower sum and the upper sum respectively of f with respect to the subdivision D). Let the numbers $\underline{I}$ and $\overline{I}$ (the lower and the upper integral respectively of f on the interval) be defined by the equalities

$$\underline{I} = \sup s(D) \qquad \text{and} \qquad \overline{I} = \inf S(D),$$

where the supremum and the infimum are taken over the set of all subdivisions of (a, b). It can be proved that $\underline{I} \leq \overline{I}$. Simple examples can be constructed where $\underline{I} = \overline{I}$, and examples can be constructed where $\underline{I} < \overline{I}$. If

$$\underline{I} = \overline{I}, \tag{3}$$

the function is said to be integrable on (a, b), and the common value of the two members of (3) is called the integral of f on (a, b) and is denoted by

$$\int_a^b f(x)\, dx. \tag{4}$$

It is seen that, in these definitions of "integrable" and "integral", the interval (a, b) can be replaced by any among the last three intervals in (1). It is also seen that a function that is continuous on a closed interval $[a, b]$, is integrable on $[a, b]$.

We want to extend these definitions to certain unbounded functions. To this end, suppose that the function f satisfies the conditions in the comments applying to Fig. 1.1 with the word "bounded" deleted. First suppose that the range of f contains no negative numbers. Let m be a positive integer, and let f_m be the function defined by

$$f_m(x) = \begin{cases} f(x), & \text{if } x \in D_f \text{ and } f(x) \leq m \\ m, & \text{if } x \in D_f \text{ and } f(x) > m. \end{cases}$$

(The curve $y = f(x)$ is "truncated" by the line $y = m$.) Suppose that $f_m(x)$ is integrable on (a, b) for every m and that the limit

$$\lim_{m \to \infty} \int_a^b f_m(x)\, dx \tag{5}$$

exists, i.e. that (5) denotes a real number. Then f is said to be integrable on (a, b), and the number (5) is called its integral on (a, b) and is denoted by (4).

Then suppose that the range of f contains negative numbers. Let the functions f^+ and f^- (the positive part and the negative part of f) be defined by

$$f^+(x) = \begin{cases} f(x), & \text{if } x \in D_f \text{ and } f(x) \geq 0 \\ 0, & \text{if } x \in D_f \text{ and } f(x) < 0, \end{cases}$$

$$f^-(x) = \begin{cases} 0, & \text{if } x \in D_f \text{ and } f(x) > 0 \\ f(x), & \text{if } x \in D_f \text{ and } f(x) \leq 0. \end{cases}$$

Suppose that the functions f^+ and $-f^-$ are both integrable on (a, b). Then f is said to be integrable on (a, b), and by its integral on (a, b), denoted by (4), is understood

$$\int_a^b f^+(x)\, dx - \int_a^b [-f^-(x)]\, dx. \tag{6}$$

We sum up our definitions in the following way. A function f is said to be *integrable* on the interval (a, b) if it satisfies the conditions given in the context for any of the formulas (3), (5) and (6). The number given by (3), (5) or (6) is called the *integral* of f on (a, b); it is denoted by (4).

Every function occurring in this text as an integrand on a finite interval is integrable in the above sense (Riemann integrable) on the interval of integration, and the operations that we perform on integrals can be proved to be legitimate. For a definition of a concept of integral (due to Lebesgue) which applies to a considerably wider class of functions than we study in this text, see [13], Chapter 10.

EXERCISE

*101. Decide for each of the following functions if it is integrable in the sense above on the interval $(0, 1)$:

a) $y = \dfrac{\sin x}{x}$

b) $y = \dfrac{\cos x}{\sqrt{x}}$

c) $y = \dfrac{e^x}{x}$

d) $y = \dfrac{1}{x} \sin \dfrac{1}{x}$

e) $f(x) = \begin{cases} 1, & \text{if } x \text{ is a rational number} \\ 0, & \text{if } x \text{ is an irrational number.} \end{cases}$

1.2 FOURIER SERIES AND FOURIER COEFFICIENTS

1.2.1 Definitions

Let a function $f(x)$ and a number $p \neq 0$ be given. If, for every x in the domain of $f(x)$, the numbers $x + p$ and $x - p$ belong to the domain and $f(x + p) = f(x)$, then the function is said to be *periodic* with the *period p*.

If p is a period of $f(x)$, then every nonzero multiple of p is a period. If two functions $f(x)$ and $g(x)$ have the period p, then their sum and product have the same period.

1.2.2 Definitions

The functions $\cos x$ and $\sin x$ both have period 2π. It follows that the functions

$\cos vx$ and $\sin vx$, v being a nonnegative integer, have period 2π. Further, every *trigonometric polynomial*, i.e. every function of the form

$$\tfrac{1}{2}a_0 + (a_1 \cos x + b_1 \sin x) + (a_2 \cos 2x + b_2 \sin 2x) + \cdots + (a_n \cos nx + b_n \sin nx),$$

or, using the sum notation,

$$\tfrac{1}{2}a_0 + \sum_{v=1}^{n} (a_v \cos vx + b_v \sin vx), \tag{1}$$

the a_v and b_v being given constants, has the period 2π. (The factor $\tfrac{1}{2}$ in the first term is introduced for convenience; see the comment on formula (6a) below.) The integer n, occurring in (1), is called the *order* of the trigonometric polynomial.

Suppose that we want to study a given function $g(x)$ of period p. We can then standardize the period to 2π by studying the function $f(x) = g(px/(2\pi))$. It may turn out that (1) is a good approximation to $f(x)$. It may turn out that a trigonometric series (to be defined below) gives a still better description of the function. Much of our work in this book will be a study of such approximations and descriptions. This study will familiarize us with some important tools of pure and applied mathematics.

1.2.3 Definitions

By a *trigonometric series* is understood a series of the form

$$\tfrac{1}{2}a_0 + \sum_{n=1}^{\infty} (a_n \cos nx + b_n \sin nx), \tag{2}$$

where x is a real variable and the coefficients $a_0, a_1, b_1, a_2, b_2, \ldots$ are real or complex numbers. Unless otherwise stated we suppose that they are real.

The nth *partial sum*, denoted by $s_n(x)$, of the series (2) is the sum (1).

Suppose there exists a nonempty subset S of the set of real numbers R, such that the series (2) is convergent, i.e. $f(x) = \lim_{n \to \infty} s_n(x)$ exists, if and only if $x \in S$. Then $f(x)$ is a function of x with domain S. The function $f(x)$ is called the *sum* of the series (2). This is written

$$f(x) = \tfrac{1}{2}a_0 + \sum_{n=1}^{\infty} (a_n \cos nx + b_n \sin nx). \tag{3}$$

As the partial sums $s_n(x)$ have the period 2π, the function $f(x)$ has the same property:

$$f(x + 2\pi) = f(x) \qquad \text{for} \quad x \in S. \tag{4}$$

In Fig. 1.2, we shall see an example of (3) and (4).

1.2.4 Auxiliary theorems

We want to express the coefficients a_n and b_n in terms of the function $f(x)$. (This is possible under suitable assumptions.) We shall then use the following equations, where m and n denote positive integers (the reader should check these equations):

$$\int_{-\pi}^{\pi} \cos mx \, dx = \int_{-\pi}^{\pi} \sin mx \, dx = 0 \tag{5a}$$

$$\int_{-\pi}^{\pi} \cos mx \cos nx \, dx = \begin{cases} \pi, & \text{if } m = n \\ 0, & \text{if } m \neq n \end{cases} \tag{5b}$$

$$\int_{-\pi}^{\pi} \sin mx \sin nx \, dx = \begin{cases} \pi, & \text{if } m = n \\ 0, & \text{if } m \neq n \end{cases} \tag{5c}$$

$$\int_{-\pi}^{\pi} \sin mx \cos nx \, dx = 0. \tag{5d}$$

The notation Z^+ is used to denote the set of all positive integers, i.e. the numbers $1, 2, 3, \ldots$. The notation N is used to denote the set of all natural numbers, i.e. the nonnegative integers $0, 1, 2, \ldots$. The notation Z is used to denote the set of all integers, i.e. the numbers $\ldots, -2, -1, 0, 1, 2, \ldots$.

We shall rely on some basic properties of uniformly convergent series. Suppose that the functions $u_n(x)$, $n \in N$, are defined on a closed interval $[a, b]$. If the series $\sum_{n=0}^{\infty} u_n(x)$ is convergent for each x in the interval, the sum $s(x)$ of the series is a function of x on the interval. The series is said to be *uniformly convergent* on $[a, b]$, if to each number $\varepsilon > 0$ there corresponds a number n_0 such that $\left| s(x) - \sum_{v=0}^{n} u_v(x) \right| < \varepsilon$ for $n > n_0$ and each x in the interval. For uniformly convergent series the following three theorems can be proved (see [13], pp. 134–138).

a) Suppose that $\sum_{n=0}^{\infty} a_n$ is a convergent series with constant positive terms and that there exists a positive number M such that $|u_n(x)| \leqslant M a_n$ for each n and all x in the interval $[a, b]$. Then the series $\sum_{n=0}^{\infty} u_n(x)$ is uniformly convergent on the interval. (This theorem is called the Weierstrass M test. The first series is called a comparison series for the second series.)

b) The sum of a uniformly convergent series of continuous functions is a continuous function.

c) A uniformly convergent series of continuous functions on a closed interval can be integrated term by term, i.e.

$$\int_a^b s(x) \, dx = \sum_{n=0}^{\infty} \int_a^b u_n(x) \, dx.$$

In Theorem (c), it should be observed that the left member of the equation has, by Theorem (b), a meaning.

1.2.5 Definitions

We now return to equation (3), and we suppose that the series in (3) is uniformly convergent on the interval $[-\pi, \pi]$. We multiply both members of (3) by, in turn, the functions $\cos nx$, $n \in N$, and $\sin nx$, $n \in Z^+$, and we integrate term by term over $[-\pi, \pi]$; this is legitimate by Theorem (c). These operations and the equations (5) give

$$a_n = \frac{1}{\pi} \int_{-\pi}^{\pi} f(x) \cos nx \, dx, \qquad n \in N, \tag{6a}$$

$$b_n = \frac{1}{\pi} \int_{-\pi}^{\pi} f(x) \sin nx \, dx, \qquad n \in Z^+. \tag{6b}$$

It should be observed that these formulas hold from $n = 0$ on and from $n = 1$ on respectively. The factor $\frac{1}{2}$ in the first term of (1) and (2) was introduced in order to make the formula (6a) hold for $n = 0$ also.

We started above from the series (2), and under certain assumptions we reached formulas (6). Now we will reverse the process.

To this end, let the function $f(x)$ be given, and let this function be integrable on the interval $(-\pi, \pi)$. Formulas (6) then define certain numbers a_n and b_n, which are called the *Fourier coefficients* of the function $f(x)$. (They are named after the French mathematician Fourier, who started this theory in the beginning of the 19th century.) The series (2) with these numbers a_n and b_n as coefficients is called the *Fourier series* of the function $f(x)$. To indicate that the function $f(x)$ has (2) as its Fourier series, the following notation

$$f(x) \sim \tfrac{1}{2}a_0 + \sum_{n=1}^{\infty} (a_n \cos nx + b_n \sin nx) \tag{7}$$

is used. The sign $\sim$ can be read : "has the Fourier series". We cannot use an equality sign in (7), because we do not know if the series converges or if, when convergent, it has the sum $f(x)$. The formation of formula (7) for a given function is called *expansion* of the function in a Fourier series.

1.2.6 Remarks

If two functions f and g are integrable on the interval $(-\pi, \pi)$ and if the values $f(x)$ and $g(x)$ are equal except at a finite number of points in the interval, then f and g have the same Fourier series.

If a function f is integrable on the interval $(-\pi, \pi)$ and periodic with the period 2π, we can substitute for the interval of integration $(-\pi, \pi)$ in (6) any interval of length 2π, for example the interval $(0, 2\pi)$.

If a (not necessarily periodic) function f is integrable on an interval of the form $(a, a + 2\pi)$, we can define numbers a_n and b_n by equations analogous to (6), using

$(a, a + 2\pi)$ as the interval of integration, and then form (2) and (7) with these numbers as coefficients. Then a_n and b_n are called the *Fourier coefficients* of $f(x)$ on $(a, a + 2\pi)$, (2) is called the *Fourier series* of $f(x)$ on $(a, a + 2\pi)$, and exhibiting (7) is called *expanding* $f(x)$ in a Fourier series on $(a, a + 2\pi)$.

We shall see in the following that Fourier series can be used to represent fairly "arbitrary" functions. Fourier series are, therefore, important in mathematics and in applications of mathematics.

1.2.7 Example
Discuss the series

$$\cos x + 3^{-2} \cos 3x + 5^{-2} \cos 5x + \cdots + (2n - 1)^{-2} \cos (2n - 1) x + \cdots$$

from the points of view discussed above.

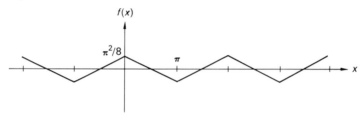

Figure 1.2

Discussion. The series is a trigonometric series, where $a_0 = 0$, $b_n = 0$ for every n, $a_n = 1/n^2$ if n is odd, and $a_n = 0$ if n is even. The series is uniformly convergent by Theorem (a); as a comparison series the series $\sum_{n=1}^{\infty} 1/n^2$ can be used. Hence it has a sum $f(x)$, defined for all real x. Further, this sum $f(x)$ is a continuous function by Theorem (b), and $f(x)$ has the given series as its Fourier series. We shall show in Exercise 120 that $f(x)$ is the periodic function of period 2π that on the interval $[-\pi, \pi]$ is equal to $(\pi^2 - 2\pi|x|)/8$ (see Fig. 1.2).

1.2.8 Example
Expand in a Fourier series the function

$$f(x) = \begin{cases} 1 \text{ for } -\pi < x < 0 \\ 2 \text{ for } \quad 0 < x < \pi \end{cases}$$

(see Figure 1.3; sometimes we denote "missing points" on a curve by small open circles).

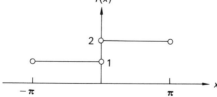

Figure 1.3

Solution. We have

$$a_0 = \frac{1}{\pi} \int_{-\pi}^{\pi} f(x)\, dx = \frac{1}{\pi} \int_{-\pi}^{0} dx + \frac{1}{\pi} \int_{0}^{\pi} 2\, dx = 3$$

and for $n > 0$

$$a_n = \frac{1}{\pi} \int_{-\pi}^{\pi} f(x) \cos nx\, dx = \frac{1}{\pi} \int_{-\pi}^{0} \cos nx\, dx + \frac{1}{\pi} \int_{0}^{\pi} 2 \cos nx\, dx = 0,$$

$$b_n = \frac{1}{\pi} \int_{-\pi}^{\pi} f(x) \sin nx\, dx = \frac{1}{\pi} \int_{-\pi}^{0} \sin nx\, dx + \frac{1}{\pi} \int_{0}^{\pi} 2 \sin nx\, dx,$$

$$b_n = \frac{2}{n\pi} \quad \text{for } n \text{ odd}, \qquad b_n = 0 \quad \text{for } n \text{ even}.$$

Hence

$$f(x) \sim \frac{3}{2} + \frac{2}{\pi}\left(\sin x + \frac{\sin 3x}{3} + \cdots + \frac{\sin(2n-1)x}{2n-1} + \cdots\right).$$

EXERCISES

102. Sketch on the interval $(-2\pi, 2\pi)$ the graph of each of the following six functions:

$\sin x$, $\cos x$, $\frac{1}{2}\sin x + (\sqrt{3}/2)\cos x$, $\sin 2x$, $\frac{1}{2}\cos 2x$, $\frac{1}{3}\sin 3x$.

103. It is known that the functions $\sin nx$, $n \in Z^+$, are continuous functions and that the series $\sum_{n=1}^{\infty} (\sin nx / n)$ is convergent for every real number x. Does it follow that the sum of the series is a continuous function?

104. It is known that

$$\sum_{n=1}^{\infty} (2n-1)^{-2} \cos(2n-1)t = (\pi^2 - 2\pi t)/8 \qquad \text{for } 0 \leqslant t \leqslant \pi.$$

What information is obtained from this equation by integrating term by term from 0 to x, when

a) $0 \leqslant x \leqslant \pi$,   b) $\pi \leqslant x \leqslant 2\pi$?

105. For which real numbers a is the series $\sum_{n=1}^{\infty} n^{-a} \cos nx$ uniformly convergent on each interval?

106. Expand the function $f(x) = x$ in a Fourier series on the interval

a) $-\pi < x < \pi$,   b) $0 < x < 2\pi$.

107. Suppose that $0 < h < \pi$ and that $f_h(x) = 1/h$ for $|x| < h$ and $= 0$ for $h < |x| < \pi$. Denote the Fourier coefficients of the function f_h by $a_n(h)$ and $b_n(h)$. Find $\lim a_n(h)$ and $\lim b_n(h)$ as $h \to 0$.

108. For which positive values of a does the function $y = |x|^{-a}$, $0 < |x| < \pi$, have a Fourier series?

*109. Prove the two assertions in 1.2.6.

110. For which real values of x does the trigonometric series

a) $\sum_{n=1}^{\infty} \cos nx$,   b) $\sum_{n=1}^{\infty} \sin nx$,

converge?

1.3 A MINIMIZING PROPERTY OF THE FOURIER COEFFICIENTS. THE RIEMANN–LEBESGUE THEOREM

Suppose that $f(x)$ is some "complicated" function, and that we want to approximate $f(x)$, on an interval (a, b), by some function $s(x)$ out of a class S of "simple" functions. One way to do this would be to determine $s(x) \in S$ such that, given a positive number p, the integral $\int_a^b |f(x) - s(x)|^p \, dx$ be minimized, provided that this integral is defined for every $s(x) \in S$, and that it has a minimum. We shall show in Problem 1.3.3 that this idea works well, when $(a, b) = (-\pi, \pi)$, $p = 2$, $f(x)$ and $f^2(x)$ are integrable on $(-\pi, \pi)$, and S is the class of trigonometric polynomials. For this we need the concept of square-integrable function and the concept of square deviation.

1.3.1 Definition

Suppose that f is a function and that (a, b) is an interval. Then f is *square integrable* on (a, b) if the functions f and f^2 are both integrable on (a, b).

1.3.2 Definition

Suppose that (a, b) is an interval and that $f(x)$ and $s(x)$ are functions such that the integral $\int_a^b [f(x) - s(x)]^2 \, dx$ is defined. Then we call this integral the *square deviation* of $s(x)$ from $f(x)$ on (a, b).

1.3.3 Problem

Suppose that the function $f(x)$ is square integrable on the interval $(-\pi, \pi)$. Let n be a given positive integer, and consider the trigonometric polynomials $t_n(x)$ of order n:

$$t_n(x) = \tfrac{1}{2}c_0 + \sum_{v=1}^{n} (c_v \cos vx + d_v \sin vx). \tag{1}$$

How shall the constants c_v and d_v be chosen in order that the square deviation of $t_n(x)$ from $f(x)$ on $(-\pi, \pi)$,

$$\int_{-\pi}^{\pi} [f(x) - t_n(x)]^2 \, dx, \tag{2}$$

be minimized?

Solution. Substitute (1) into (2), expand the square, integrate term by term, use formulas (5) and (6) in Section 1.2.4 and Definition 1.2.5, and divide both members by π. We then get

$$\frac{1}{\pi}\int_{-\pi}^{\pi}[f(x)-t_n(x)]^2\,dx = \frac{1}{\pi}\int_{-\pi}^{\pi}f^2(x)\,dx + \tfrac{1}{2}c_0^2 + c_1^2 + d_1^2 + \dots$$
$$+ c_n^2 + d_n^2 - a_0c_0 - 2a_1c_1 - 2b_1d_1 - \dots - 2a_nc_n - 2b_nd_n,$$

where the a_v and the b_v are Fourier coefficients of $f(x)$. Add and subtract the sum $\tfrac{1}{2}a_0^2 + \sum_{v=1}^{n}(a_v^2 + b_v^2)$ to the right of this equation. We get

$$\frac{1}{\pi}\int_{-\pi}^{\pi}[f(x)-t_n(x)]^2\,dx = \frac{1}{\pi}\int_{-\pi}^{\pi}f^2(x)\,dx - \tfrac{1}{2}a_0^2 - \sum_{v=1}^{n}(a_v^2 + b_v^2)$$
$$+ \tfrac{1}{2}(c_0 - a_0)^2 + \sum_{v=1}^{n}[(c_v - a_v)^2 + (d_v - b_v)^2].$$

Here the terms in the last line are nonnegative, and they become equal to zero if we choose $c_v = a_v$ for $v = 0, \dots, n$ and $d_v = b_v$ for $v = 1, \dots, n$. Hence the integral (2) becomes a minimum if, as coefficients in $t_n(x)$, we choose the corresponding Fourier coefficients of the function $f(x)$. Thereby our problem is solved.

1.3.4 Remark

We can now deduce an important formula called the Bessel inequality. Setting $c_v = a_v$ and $d_v = b_v$ in the above equation we obtain

$$\frac{1}{\pi}\int_{-\pi}^{\pi}[f(x)-s_n(x)]^2\,dx = \frac{1}{\pi}\int_{-\pi}^{\pi}f^2(x)\,dx - \tfrac{1}{2}a_0^2 - \sum_{v=1}^{n}(a_v^2 + b_v^2), \qquad (3)$$

where, as earlier, $s_n(x)$ denotes the nth partial sum of the Fourier series of $f(x)$. Here the left member is nonnegative, whence

$$\tfrac{1}{2}a_0^2 + \sum_{v=1}^{n}(a_v^2 + b_v^2) \leqslant \frac{1}{\pi}\int_{-\pi}^{\pi}f^2(x)\,dx.$$

As this formula holds for every n, we have

$$\tfrac{1}{2}a_0^2 + \sum_{n=1}^{\infty}(a_n^2 + b_n^2) \leqslant \frac{1}{\pi}\int_{-\pi}^{\pi}f^2(x)\,dx. \qquad (4)$$

Formula (4) is called the *Bessel inequality*. It states, roughly, that a sum of squares of Fourier coefficients of a function $f(x)$, square integrable on $(-\pi, \pi)$, cannot be arbitrarily large but is bounded above by a certain number that is determined by the function.

It follows from the Bessel inequality that the Fourier coefficients of a function $f(x)$, square integrable on $(-\pi, \pi)$, tend to zero as $n \to \infty$. This property is a special case of the following theorem, which is called the Riemann–Lebesgue theorem. (We prove it only for functions that are integrable in the sense of Section 1.1. It can be proved for Lebesgue-integrable functions.) The variable t that occurs in the statement of the theorem takes arbitrary real values (while the variable n in the definition of Fourier coefficients takes only certain integral values). In the proof we assume that a certain function $\phi(x)$ is *piecewise constant*; this means that, given any interval (a, b) in the domain of $\phi(x)$, there exists a subdivision of (a, b) such that $\phi(x)$ is constant between any two consecutive points of the subdivision. (The reader can, by suitable completion of Fig. 1.1, exhibit the graph of a function that is piecewise constant on (a, b) and whose integral on (a, b) is e.g. an upper sum.)

1.3.5   Theorem   (the Riemann–Lebesgue theorem). *Suppose that the function* $f(x)$ *is integrable on the interval* (a, b). *Then*

$$\lim_{t \to \infty} \int_a^b f(x) \cos tx \, dx = 0 \qquad and \qquad \lim_{t \to \infty} \int_a^b f(x) \sin tx \, dx = 0. \tag{5}$$

Proof. Suppose that $\varepsilon > 0$. From formulas (5) and (6) of Definition 1.1.3 it is seen that there exists a bounded function $f_m(x)$, integrable on (a, b), such that

$$\int_a^b |f(x) - f_m(x)| \, dx < \frac{\varepsilon}{3}.$$

It follows that, for every real number t,

(a) $$\left| \int_a^b f(x) \cos tx \, dx - \int_a^b f_m(x) \cos tx \, dx \right| < \frac{\varepsilon}{3}.$$

Further, by formula (3) of the same definition, there exists a piecewise constant function $\phi(x)$, defined on (a, b), such that

$$\int_a^b |f_m(x) - \phi(x)| \, dx < \frac{\varepsilon}{3}.$$

Hence, for every real number t,

(b) $$\left| \int_a^b f_m(x) \cos tx \, dx - \int_a^b \phi(x) \cos tx \, dx \right| < \frac{\varepsilon}{3}.$$

There is a subdivision of the interval (a, b), consisting of a strictly increasing finite sequence of numbers $x_0, x_1, \ldots, x_n$, such that $\phi(x)$ takes a constant value c_v in each subinterval (x_{v-1}, x_v). Then, for $t > 0$,

$$\int_a^b \phi(x) \cos tx \, dx = \sum_{v=1}^n c_v \int_{x_{v-1}}^{x_v} \cos tx \, dx = \frac{1}{t} \sum_{v=1}^n c_v (\sin tx_v - \sin tx_{v-1}),$$

and hence

$$\left| \int_a^b \phi(x) \cos tx \, dx \right| \leqslant \frac{2}{t} \sum_{v=1}^n |c_v|.$$

This shows that, for $t > t_0 = \dfrac{6}{\varepsilon} \displaystyle\sum_{v=1}^n |c_v|$,

(c) $$\left| \int_a^b \phi(x) \cos tx \, dx \right| < \frac{\varepsilon}{3}.$$

By (a), (b) and (c) there holds for $t > t_0$

(d) $$\left| \int_a^b f(x) \cos tx \, dx \right| < \varepsilon.$$

This gives the first equation (5), and the second is proved analogously.

EXERCISES

111. Are the functions
     (a) $x^{-1/4}$,　　　　　　　　　　　　　(b) $x^{-1/2}$,

     (c) $f(x) = \begin{cases} +1, & \text{if } x \text{ is a rational number,} \\ -1, & \text{if } x \text{ is an irrational number,} \end{cases}$

     square integrable on the interval $(0, 1)$?

112. Does there exist any function, integrable on the interval $(-\pi, \pi)$, that has the series $\sum_{n=1}^{\infty} \sin nx$ as its Fourier series?

113. Suppose that the function $f(x)$ is a trigonometric polynomial. Can the sign $\leqslant$ in the Bessel inequality for $f(x)$ be replaced by an equality sign?

114. For which real numbers a, b, c is the integral
     $$\int_{-\pi}^{\pi} (x^2 - a - b \cos x - c \sin x)^2 \, dx$$
     a minimum?

*115. Suppose that the functions $f(x)$ and $g(x)$ are square integrable on an interval (a, b). Show that their product $f(x) \, g(x)$ is integrable on the same interval.
     Hint (complete the following sketch to a proof). First suppose that $f(x) \geqslant 1$ and $g(x) \geqslant 1$ on (a, b). Let index m denote truncation of a function as in Definition 1.1.3. Then the function $f_m g_m$ is integrable on (a, b). Observe that

     $$(fg)_m = (f_m g_m)_m.$$

     It follows that $(fg)_m$ is integrable on (a, b). We have that

     $$(fg)_m(x) \leqslant f(x)g(x) \leqslant \tfrac{1}{2}[f^2(x) + g^2(x)] \text{ on } (a, b).$$

     Hence

     $$\lim_{m \to \infty} \int_a^b (fg)_m(x) \, dx$$

     exists finite. Therefore fg is integrable on (a, b). Then suppose that $f(x) \geqslant 0$ and $g(x) \geqslant 0$ on (a, b). The equation

     $$fg = (f+1)(g+1) - f - g - 1$$

     and the result just obtained show that fg is integrable on (a, b). Finally treat the case that f and g take negative values.

1.4 CONVERGENCE OF FOURIER SERIES

We shall state and prove a theorem on convergence of Fourier series. To do this we need a lemma, where the nth partial sum of a Fourier series is expressed by a certain integral, called the Dirichlet integral.

1.4.1 Lemma (giving the Dirichlet integral). *Suppose that the function $f(x)$ is integrable on the interval $(-\pi, \pi)$ and periodic with period 2π. Then the nth partial sum $s_n(x)$ of its Fourier series satisfies the equation*

$$s_n(x) = \frac{1}{2\pi} \int_{-\pi}^{\pi} f(x + u) \frac{\sin (n + \frac{1}{2})u}{\sin \frac{1}{2}u} \, du. \tag{1}$$

Proof. Formulas (1) and (6) of Definition 1.2.2 and Definition 1.2.5 give

$$s_n(x) = \frac{1}{2\pi} \int_{-\pi}^{\pi} f(t) \, dt + \sum_{v=1}^{n} \left[(\cos vx) \frac{1}{\pi} \int_{-\pi}^{\pi} f(t) \cos vt \, dt \right.$$

$$\left. + (\sin vx) \frac{1}{\pi} \int_{-\pi}^{\pi} f(t) \sin vt \, dt \right] = \frac{1}{\pi} \int_{-\pi}^{\pi} f(t) \left[\frac{1}{2} + \sum_{v=1}^{n} \cos v(t - x) \right] dt.$$

Further (it is left to the reader to verify this equation, e.g. by applying the formula for the sum of a finite geometric sequence to $\frac{1}{2} \sum_{v=-n}^{n} e^{ivx}$)

$$\frac{1}{2} + \sum_{v=1}^{n} \cos vx = \frac{\sin (n + \frac{1}{2})x}{2 \sin \frac{1}{2}x}. \tag{2}$$

Hence

$$s_n(x) = \frac{1}{2\pi} \int_{-\pi}^{\pi} f(t) \frac{\sin (n + \frac{1}{2})(t - x)}{\sin \frac{1}{2}(t - x)} \, dt.$$

The substitution $t - x = u$ gives

$$s_n(x) = \frac{1}{2\pi} \int_{-\pi-x}^{\pi-x} f(x + u) \frac{\sin (n + \frac{1}{2})u}{\sin \frac{1}{2}u} \, du.$$

As the integrand is periodic with period 2π, the limits $-\pi - x$ and $\pi - x$ can be replaced by $-\pi$ and π to give the integral in (1).

If a function f has a left limit and a right limit at the point a, these limits are denoted by $f(a^-)$ and $f(a^+)$ respectively.

A function f is said to be *piecewise continuously differentiable* on an interval (a, b), if there is a subdivision of the interval (a, b), consisting of a strictly increasing finite sequence of numbers $x_0, x_1, \ldots, x_n$ such that, for $v = 1, \ldots, n$, the function g_v defined by

$$g_v(x_{v-1}) = f(x_{v-1}^+), \qquad g_v(x) = f(x) \quad \text{for} \quad x_{v-1} < x < x_v, \qquad g_v(x_v) = f(x_v^-)$$

has a continuous derivative on the closed interval $[x_{v-1}, x_v]$.

We introduce a class of functions that we call the class D after Dirichlet.

1.4.2 Definition

We shall say that a function $f(x)$ belongs to the *class D* if it has the following four properties:

1. $f(x)$ has the domain $(-\infty, \infty)$
2. $f(x)$ is periodic with the period 2π

3. $f(x)$ is piecewise continuously differentiable on $(-\pi, \pi)$
4. $f(x) = \frac{1}{2}[f(x^+) + f(x^-)]$ for each real number x.

1.4.3 Remarks

Suppose that the function $f(x)$ belongs to the class D. Property 3 then implies that $f(x)$ is integrable on $(-\pi, \pi)$, whence it has a Fourier series. Properties 2 and 3 show that $f(x)$ has at most finitely many discontinuities in each finite interval and that the limits $f(x^+)$ and $f(x^-)$ exist for every real number x. Property 4 is that the value at any point x is equal to the arithmetic mean of the right and left limits.

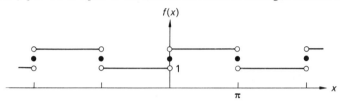

Figure 1.4

For example, the function $f(x)$ of Example 1.2.8 can be extended to a function of the class D by the three requirements: $f(0) = 3/2$, $f(\pi) = 3/2$, and $f(x + 2\pi) = f(x)$ for every real number x. (A function obtained by following the third requirement is called the *periodic extension* of the given function.) Figure 1.4 shows the function obtained; certain points of the graph of this function are denoted by dark circular disks. One more example: The function of Figure 1.2 also belongs to the class D.

We can now give a theorem on convergence of Fourier series.

1.4.4 Theorem. *If the function $f(x)$ belongs to the class D, its Fourier series is convergent for every real number x, and the Fourier series has the sum $f(x)$ for every real number x.*

Proof. Integration of the two members of (2) from 0 to π gives

$$1 = \frac{1}{\pi}\int_0^\pi \frac{\sin(n + \frac{1}{2})u}{\sin \frac{1}{2}u}\, du. \tag{3}$$

Fix x. Then (3) shows that

$$\frac{1}{\pi}\int_0^\pi f(x + u)\frac{\sin(n + \frac{1}{2})u}{\sin \frac{1}{2}u}\, du - f(x^+)$$

$$= \frac{1}{\pi}\int_0^\pi \frac{f(x + u) - f(x^+)}{\sin \frac{1}{2}u}\sin(n + \frac{1}{2})u\, du$$

$$= \frac{1}{\pi}\int_0^\pi \frac{f(x + u) - f(x^+)}{u}\frac{u}{\sin \frac{1}{2}u}\sin(n + \frac{1}{2})u\, du.$$

By Property 3 in Definition 1.4.2, $[f(x + u) - f(x^+)]/u$ tends to a limit as $u \to 0^+$. Hence the last integrand is the product of $\sin (n + \frac{1}{2})u$ and a function that is integrable on $(0, \pi)$. The Riemann–Lebesgue theorem (1.3.5) then shows that

$$\frac{1}{\pi} \int_0^\pi f(x + u) \frac{\sin (n + \frac{1}{2})u}{\sin \frac{1}{2}u} \, du - f(x^+) \to 0 \qquad \text{as} \quad n \to \infty.$$

Analogously it is seen that

$$\frac{1}{\pi} \int_{-\pi}^0 f(x + u) \frac{\sin (n + \frac{1}{2})u}{\sin \frac{1}{2}u} \, du - f(x^-) \to 0 \qquad \text{as} \quad n \to \infty.$$

These two properties and Lemma 1.4.1 show that

$$s_n(x) - \tfrac{1}{2} \left[f(x^+) + f(x^-) \right] \to 0 \qquad \text{as} \quad n \to \infty.$$

The assertion now follows.

Change the assumptions in the above theorem in the following way: $f(x)$ need not be defined at certain points, at most finitely many in every finite interval; $f(x)$ is not required to have property 4 of Definition 1.4.2. The following corollary is then obtained.

1.4.5 Corollary. *If the function $f(x)$ is piecewise continuously differentiable on the interval $(-\pi, \pi)$ and periodic with the period 2π, its Fourier series is convergent for every real number x, and the Fourier series has the sum $\frac{1}{2} \left[f(x^+) + f(x^-) \right]$ for every real number x.*

The existing theory of convergence for Fourier series is quite extensive (see [19]). Parts of this theory are difficult and offer some astonishing results. For example, a Fourier series can be divergent for every x, and a Fourier series can diverge at a point of continuity. In fact, it was not known until quite recently whether the Fourier series of a continuous function has to converge anywhere. Carleson has settled this question (in 1966) by showing that the Fourier series of any square-integrable function is convergent almost everywhere (for an explanation of the term "almost everywhere", see [13], p. 243).

We have above, for practical reasons, given a "global" theorem on convergence of Fourier series, i.e. a theorem on convergence for every x. It is also possible to give "local" theorems, i.e. theorems on convergence for a particular x. Such a theorem will be given in Exercise 121.

1.4.6 Example

By Exercise 106 and Theorem 1.4.4, the series $2 \sum_{n=1}^{\infty} [(-1)^{n+1} \sin nx/n]$ has the sum x for $-\pi < x < \pi$ and the sum zero when x is a multiple of π. Figure 1.5 shows the partial sums $s_2(x)$, $s_3(x)$, $s_4(x)$ and the sum of this series on the closed interval $[0, \pi]$. The figure illustrates the convergence of the series to its sum.

1.4.7 Remarks

When a given function $f(x)$ is to be expanded in a Fourier series, it is convenient to start by examining whether the function, perhaps after periodic continuation, satisfies the hypotheses of Theorem 1.4.4 or its corollary. If so, the theorem or

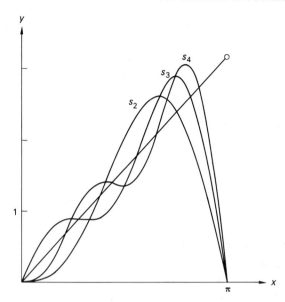

Figure 1.5

the corollary gives the sum of the series for all real numbers x. Thus, Figure 1.4 shows the sum of the Fourier series of Example 1.2.8.

Further, it should be observed that the expressions for the Fourier coefficients can be simplified, if $f(x)$ is an odd function or an even function. If the function $f(x)$ is *odd*, i.e. if $f(-x) = -f(x)$ for all $x \in D_f$, there holds for its Fourier coefficients:

$$a_n = 0, \tag{4a}$$

$$b_n = \frac{2}{\pi} \int_0^\pi f(x) \sin nx \, dx, \tag{4b}$$

for the integrand of formula (6a) in Definition 1.2.5 is odd and the integrand of (6b) is even. Analogously, if $f(x)$ is *even*, i.e. if $f(-x) = f(x)$ for all $x \in D_f$, then

$$a_n = \frac{2}{\pi} \int_0^\pi f(x) \cos nx \, dx, \tag{5a}$$

$$b_n = 0, \tag{5b}$$

for the integrand of (6a) is even and the integrand of (6b) is odd. The function $f(x)$ in Example 1.2.8 is neither even nor odd; however, $f(x) - \frac{1}{2}a_0$ is an odd function, and hence $a_n = 0$ for $n > 0$.

EXERCISES

For each of Exercises 116–119 the reader should draw a figure, showing the sum of the Fourier series for $-2\pi \leqslant x \leqslant 2\pi$.

116. Expand the function $f(x) = (\pi - x)/2, 0 < x < 2\pi$, in a Fourier series.

117. a) Expand in a Fourier series the function f defined by

$f(x) = 0$ for $-\pi < x < 0$, $f(x) = x/\pi$ for $0 < x < \pi$.

b) Use the result to sum the series $\sum_{n=1}^{\infty} (2n - 1)^{-2}$.

118. a) Expand the function $f(x) = x^2, 0 < x < 2\pi$, in a Fourier series.

b) Use the result to sum the series $\sum_{n=1}^{\infty} n^{-2}$.

119. Expand the function $f(x) = \cosh x$, $-\pi < x < \pi$, in a Fourier series.

120. Prove the assertion in the comment to Fig. 1.2.

*121. Prove the following theorem (the Dini theorem): Suppose that the function $f(x)$ is integrable on the interval $(-\pi, \pi)$ and periodic with period 2π and that a and s are real numbers. Set $\phi(u) = f(a + u) + f(a - u) - 2s$. Suppose that the function $\phi(u)/u$ is integrable on some right neighborhood of the point $u = 0$. Then the Fourier series of the function $f(x)$ converges at the point $x = a$ to the sum s.

1.5 THE PARSEVAL FORMULA

The inequality sign in the Bessel inequality (4) in Remark 1.3.4 can, in fact, be replaced by an equality sign. Thereby we get formula (1) below, called the Parseval formula. For its proof we need two lemmas.

1.5.1 Lemma. *If the function $f(x)$ is periodic with period 2π and has a continuous second derivative, then the Fourier coefficients a_n and b_n of $f(x)$ have the properties:*

$$\lim_{n \to \infty} n^2 a_n = 0 \quad \text{and} \quad \lim_{n \to \infty} n^2 b_n = 0.$$

Proof. Integration by parts twice gives for $n > 0$

$$a_n = \frac{1}{\pi} \int_{-\pi}^{\pi} f(x) \cos nx \, dx$$

$$= -\frac{1}{\pi n} \int_{-\pi}^{\pi} f'(x) \sin nx \, dx$$

$$= -\frac{1}{\pi n^2} \int_{-\pi}^{\pi} f''(x) \cos nx \, dx.$$

Here $-1/n^2$ is multiplied by the nth Fourier cosine coefficient of the function $f''(x)$, which function is integrable on $(-\pi, \pi)$. The Riemann–Lebesgue theorem 1.3.5 then gives the first assertion. The second assertion is proved analogously.

1.5.2 Remark

Suppose that $P_1 = (x_1, y_1)$ and $P_2 = (x_2, y_2)$, where $x_1 < x_2$ and $y_1 \neq y_2$ are two points in a coordinate system (x, y). Then there exists a function $y = f(x)$ such that $f(x)$ is monotonic on the closed interval $[x_1, x_2]$, $f''(x)$ exists and is continuous

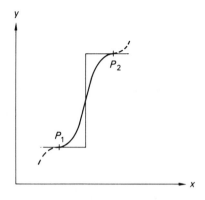

Figure 1.6

on $[x_1, x_2]$, P_1 and P_2 are points of inflection of the curve $y = f(x)$, and this curve has horizontal tangents at P_1 and P_2 (see Figure 1.6 and Exercise 126).

1.5.3 Lemma. *Suppose that the function f(x) is integrable on the interval* $(-\pi, \pi)$, *that* $\varepsilon > 0$, *that m is a positive integer, and that* $|f(x)| \leqslant m$ *for every x in* $(-\pi, \pi)$. *Then there is a positive integer n and a trigonometric polynomial* $t_n(x)$ *of order n such that*

$$\int_{-\pi}^{\pi} |f(x) - t_n(x)|\, dx < \varepsilon$$

and that for every real number x

$$|t_n(x)| < m + 1.$$

Proof. By Definition 1.1.3 there exists a piecewise constant function $\phi(x)$ such that

(a) $$\int_{-\pi}^{\pi} |f(x) - \phi(x)|\, dx < \frac{\varepsilon}{3}.$$

(We now use Remark 1.5.2, as we "round off the vertices" of the curve $y = \phi(x)$ as in Figure 1.6.) The remark shows that there is a function $g(x)$, satisfying the assumptions of Lemma 1.5.1, such that

(b) $$\int_{-\pi}^{\pi} |\phi(x) - g(x)|\, dx < \frac{\varepsilon}{3}.$$

By Lemma 1.5.1. and Theorem 1.2.4 (a) with the series $\sum_{n=1}^{\infty} n^{-2}$ as a comparison series, the Fourier series of $g(x)$ is uniformly convergent. By Theorem 1.4.4 this Fourier series has the sum $g(x)$. Hence, for n sufficiently large, its nth partial sum $t_n(x)$ satisfies the inequality

$$|g(x) - t_n(x)| < \frac{\varepsilon}{6\pi}$$

for all real x. Thus

(c) $$\int_{-\pi}^{\pi} |g(x) - t_n(x)| \, dx < \frac{\varepsilon}{3}.$$

The inequalities (a), (b), (c) give the first assertion. The above approximations can obviously be so arranged that $|\phi(x)| \leqslant m$, that $|g(x)| \leqslant m$ and, as the Fourier series of $g(x)$ converges uniformly to $g(x)$, $|t_n(x)| < m + 1$ for every real x. This is the second assertion.

1.5.4 Theorem (giving the Parseval formula). *If the function $f(x)$ is square integrable on the interval $(-\pi, \pi)$ and has the Fourier coefficients a_n and b_n, then*

$$\tfrac{1}{2}a_0^2 + \sum_{n=1}^{\infty} (a_n^2 + b_n^2) = \frac{1}{\pi} \int_{-\pi}^{\pi} f^2(x) \, dx. \tag{1}$$

Proof. Suppose that $\varepsilon > 0$. Let m be a positive integer. Denote by f_m the function for which the value $f_m(x) = m$ if $f(x) > m$, $f_m(x) = f(x)$ if $|f(x)| \leqslant m$ and $f_m(x) = -m$ if $f(x) < -m$. (The curve $y = f(x)$ is "truncated" by the lines $y = m$ and $y = -m$). Suppose that m is so large that

$$\int_{-\pi}^{\pi} [f^2(x) - f_m^2(x)] \, dx < \frac{\varepsilon}{4}$$

(by Definitions 1.1.3 and 1.3.1 there exists such an m). As $f(x)$ and $f_m(x)$ have the same sign if they are not both zero,

$$\int_{-\pi}^{\pi} [f(x) - f_m(x)]^2 \, dx \leqslant \int_{-\pi}^{\pi} |f(x) + f_m(x)| \, |f(x) - f_m(x)| \, dx$$

$$= \int_{-\pi}^{\pi} [f^2(x) - f_m^2(x)] \, dx,$$

and hence

(d) $$\int_{-\pi}^{\pi} [f(x) - f_m(x)]^2 \, dx < \frac{\varepsilon}{4}.$$

By Lemma 1.5.3 there is a trigonometric polynomial $t_n(x)$ such that for every real number x

$$|t_n(x)| < m + 1 \qquad \text{and} \qquad \int_{-\pi}^{\pi} |f_m(x) - t_n(x)| \, dx < \frac{\varepsilon}{4(2m + 1)}.$$

It follows that

$$\int_{-\pi}^{\pi} [f_m(x) - t_n(x)]^2 \, dx \leqslant (2m + 1) \int_{-\pi}^{\pi} |f_m(x) - t_n(x)| \, dx,$$

and further that

(e) $$\int_{-\pi}^{\pi} [f_m(x) - t_n(x)]^2 \, dx < \frac{\varepsilon}{4}.$$

The inequality

(f) $$(a + b)^2 \leqslant 2(a^2 + b^2),$$

which holds for arbitrary real numbers a and b (Exercise 122), and the inequalities (d) and (e) show that

(g) $$\int_{-\pi}^{\pi} [f(x) - t_n(x)]^2 \, dx < \varepsilon.$$

Here replace $t_n(x)$ by the nth partial sum $s_n(x)$ of the Fourier series of $f(x)$. The solution of Problem 1.3.3 shows that the square deviation from $f(x)$ on $(-\pi, \pi)$ does not increase. Hence

(h) $$\int_{-\pi}^{\pi} [f(x) - s_n(x)]^2 \, dx < \varepsilon.$$

By formula (3) in Remark 1.3.4, index n in the inequality (h) can be replaced by any larger index. Hence

$$\int_{-\pi}^{\pi} [f(x) - s_n(x)]^2 \, dx \to 0 \qquad \text{as} \quad n \to \infty. \tag{2}$$

Formula (3) in Remark 1.3.4 now yields the assertion.

The Parseval formula can be "proved" by replacing $f(x)$ in the right member of (1) by its Fourier series, "expanding the square", and integrating term by term. These operations, however, are not legitimate, except under special conditions.

Has Theorem 1.5.4 a converse in the sense that, given a sequence of numbers $a_0, a_1, b_1,$ $\ldots, a_n, b_n, \ldots$ such that the series in the left member of (1) is convergent, there exists a square-integrable function $f(x)$ having these numbers as its Fourier coefficients? The answer to this question is affirmative in the class of Lebesgue-integrable functions (see [13], p. 255) but not in the class of Riemann-integrable functions. This converse of Theorem 1.5.4 is called the Riesz–Fischer theorem.

1.5.5 Example

Exercise 106 shows that the function $f(x) = x/2$, $-\pi < x < \pi$, has the Fourier series

$$\sum_{n=1}^{\infty} (-1)^{n+1} \frac{\sin nx}{n}.$$

By Theorem 1.5.4, the series $\sum_{n=1}^{\infty} n^{-2}$ then has the sum

$$\frac{1}{\pi} \int_{-\pi}^{\pi} \left(\frac{x}{2}\right)^2 \, dx = \frac{\pi^2}{6}.$$

Hence we have the result

$$1 + \frac{1}{4} + \frac{1}{9} + \cdots + \frac{1}{n^2} + \cdots = \frac{\pi^2}{6}. \tag{3}$$

EXERCISES

122. a) Prove the inequality (f) above.
   b) Is the inequality (f) "best possible" in the sense that the factor 2 in the right member cannot be replaced by any smaller number?
   Hint. First show that $(a + b)^2 + (a - b)^2 = 2(a^2 + b^2)$.

123. Expand the function $(1 + \tan^2 x)^{-1}$ in a Fourier series. What result is obtained by applying the Parseval formula to this function?

124. Expand the function $f(x) = |x|$, $|x| < \pi$, in a Fourier series. Sum the series $\sum_{n=1}^{\infty}(2n - 1)^{-4}$.

125. Expand the function $x^3 - \pi^2 x$, $|x| < \pi$, in a Fourier series. Sum the series $\sum_{n=1}^{\infty} n^{-6}$.

*126. Show that the function

$$y = b + k(x - a) + \frac{kh}{\pi} \sin \frac{\pi(x - a)}{h},$$

   where

   $$a = \tfrac{1}{2}(x_1 + x_2), \qquad b = \tfrac{1}{2}(y_1 + y_2), \qquad h = \tfrac{1}{2}(x_2 - x_1), \qquad k = \frac{y_2 - y_1}{x_2 - x_1},$$

   has the properties stated in Remark 1.5.2.

127. Given a positive number ε and a function $f(x)$, square integrable on an interval (a, b) with $b \leqslant a + 2\pi$, prove the existence of a trigonometric polynomial whose square deviation from $f(x)$ on (a, b) is smaller than ε.

128. The function $f(x)$ has a continuous derivative on the interval $[0, \pi]$. Further $f(0) = f(\pi) = 0$. Show that

$$\int_0^\pi f^2(x)\, dx \leqslant \int_0^\pi f'^2(x)\, dx.$$

   For what functions does equality hold?

129. Prove the Weierstrass approximation theorem: Suppose that $\varepsilon > 0$ and that the function $f(x)$ is continuous on the closed interval $[a, b]$. Then there exists a polynomial $p(x)$ such that $|f(x) - p(x)| < \varepsilon$ on $[a, b]$.
   Hint. Set $g(y) = a + (b - a)\,y/\pi = x$. Define a function $h(y)$, continuous on $(-\infty, \infty)$ and of period 2π, such that $h(y) = f[g(y)]$ on $[0, \pi]$. By a proof similar to that of Lemma 1.5.3, prove the existence of a trigonometric polynomial $t(y)$ such that $|h(y) - t(y)| < \varepsilon/2$ for all real numbers y. The Taylor series $\sum_{n=0}^{\infty} c_n y^n$ of $t(y)$ has a partial sum $s(y)$ such that $|t(y) - s(y)| < \varepsilon/2$ on $[0, \pi]$. Show that $s[g^{-1}(x)]$ satisfies the assertion.

1.6 DETERMINATION OF THE SUM OF CERTAIN TRIGONOMETRIC SERIES

In this section (1.6), we assume that the reader has some knowledge of functions $f(z)$ analytic in a domain of the complex plane (see[1], especially pp. 38–42 and 177).

1.6.1 Generalities

Suppose that S is a set in the complex plane which is either empty or consists of finitely many points $z_1, \ldots, z_n$ on the unit circle $|z| = 1$, that $f(z)$ is a function analytic for $|z| \leqslant 1$ except at points belonging to S, that $\sum_{n=0}^{\infty} c_n z^n$ is the Taylor series of $f(z)$, and that this series is convergent when $|z| = 1$ and $z \notin S$. Then for x real and $e^{ix} \notin S$,

$$\sum_{n=0}^{\infty} c_n e^{inx} = f(e^{ix}). \tag{1}$$

Separating into real and imaginary parts, we get for $e^{ix} \notin S$

$$c_0 + \sum_{n=1}^{\infty} c_n \cos nx = \operatorname{Re} f(e^{ix}) \quad \text{and} \quad \sum_{n=1}^{\infty} c_n \sin nx = \operatorname{Im} f(e^{ix}). \tag{2}$$

Then, if the coefficients c_n of a cosine or sine series are given and if we know a function $f(z)$, satisfying the above conditions and having the c_n as its Taylor coefficients, we can use (2) to sum the given series.

1.6.2 Example

Sum the two series

$$1 + \sum_{n=1}^{\infty} \frac{\cos nx}{2^n} \quad \text{and} \quad \sum_{n=1}^{\infty} \frac{\sin nx}{2^n}.$$

Solution. By 1.6.1,

$$1 + \sum_{n=1}^{\infty} \frac{\cos nx}{2^n} + i \sum_{n=1}^{\infty} \frac{\sin nx}{2^n} = \sum_{n=0}^{\infty} \frac{e^{inx}}{2^n} = \sum_{n=0}^{\infty} \left(\frac{z}{2}\right)^n$$

$$= \frac{1}{1 - \dfrac{z}{2}} = \frac{2}{2 - z} = \frac{2}{2 - \cos x - i \sin x} = \frac{2 - \cos x + i \sin x}{2 - \cos x + i \sin x}.$$

Hence, the sums of the series are

$$\frac{4 - 2\cos x}{5 - 4\cos x} \quad \text{and} \quad \frac{2\sin x}{5 - 4\cos x}$$

respectively.

1.6.3 Example

Find the functions f and g defined by

$$f(x) = \sum_{n=1}^{\infty} \frac{\cos nx}{n} \quad \text{and} \quad g(x) = \sum_{n=1}^{\infty} \frac{\sin nx}{n}.$$

Solution. Consider for $|z| < 1$ the branch of log $(1 + z)$ that takes the value zero at the origin. It has the Taylor series $\sum_{n=1}^{\infty}(-1)^{n-1}z^n/n$ (see [1], p. 178). For $0 < x < 2\pi$, by Section 1.6.1

$$f(x) + i\, g(x) = \sum_{n=1}^{\infty} e^{inx}/n = \sum_{n=1}^{\infty} z^n/n = -\log\,(1 - z)$$

$$= -\log\,(1 - \cos x - i \sin x)$$

$$= -\log\left(2\sin^2\frac{x}{2} - 2i \sin\frac{x}{2}\cos\frac{x}{2}\right)$$

$$= -\log\left[2\sin\frac{x}{2}\left(\sin\frac{x}{2} - i\cos\frac{x}{2}\right)\right]$$

$$= -\log\left(2\sin\frac{x}{2}\right) - \log\left(\cos\frac{x-\pi}{2} + i\sin\frac{x-\pi}{2}\right)$$

$$= -\log\left(2\sin\frac{x}{2}\right) - i\frac{x-\pi}{2},$$

$$f(x) = -\log\left(2\sin\frac{x}{2}\right) \quad\text{and}\quad g(x) = \frac{\pi-x}{2}.$$

When x is a multiple of 2π, the first series is divergent and the second series has the sum zero. For values of x outside the interval $[0, 2\pi]$, the sums of the series are readily obtained, as they are periodic functions with period 2π (see Fig. 1.7).

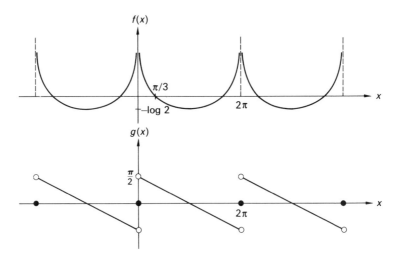

Figure 1.7

EXERCISES

130. Sum the series

$$1 + \sum_{n=1}^{\infty} \frac{\cos nx}{n!} \text{ and } \sum_{n=1}^{\infty} \frac{\sin nx}{n!}.$$

131. For $-\pi < x < \pi$, sum the series

$$\sum_{n=1}^{\infty} (-1)^{n+1} \frac{\cos nx}{n} \quad \text{and} \quad \sum_{n=1}^{\infty} (-1)^{n+1} \frac{\sin nx}{n}.$$

132. Show that the two series in Example 1.6.3 are, in fact, the Fourier series of the functions $f(x)$ and $g(x)$ respectively. (This result will be used in Exercise 227.)

    Hint. The following chain of equalities, obtained by the substitutions $x = 2t$ and $t = \pi - u$, may be needed:

$$\int_0^{\pi} \log (2 \sin \tfrac{x}{2}) dx = \pi \log 2 + \int_0^{\pi} \log \sin \tfrac{x}{2} dx$$

$$= \pi \log 2 + 2 \int_0^{\pi/2} \log \sin t \, dt$$

$$= \pi \log 2 + 2 \int_0^{\pi/2} \log (2 \sin \tfrac{t}{2} \cos \tfrac{t}{2}) dt$$

$$= 2\pi \log 2 + 2 \int_0^{\pi/2} \log \sin \tfrac{t}{2} dt + 2 \int_{\pi/2}^{\pi} \log \sin \tfrac{u}{2} du$$

$$= 2 \int_0^{\pi} \log (2 \sin \tfrac{x}{2}) dx.$$

It may also be necessary to compute the integral $\int_0^{\pi} \cot t \sin 2nt \, dt$. Show that it is equal to $2 \int_0^{\pi} \cos t [\cos (2n-1)t + \cos (2n-3)t + \ldots + \cos t] dt = 2 \int_0^{\pi} \cos t \cos t \, dt$.

CHAPTER 2

ORTHOGONAL SYSTEMS

2.1 INTEGRATION OF COMPLEX-VALUED FUNCTIONS OF A REAL VARIABLE

We assume that the reader has some acquaintance with complex numbers and complex-valued functions. In this section (2.1) we give a survey of some concepts to be used. We also introduce a convention.

2.1.1 Definitions

A *complex-valued function of a real variable* is a set of ordered pairs (x, w) such that each first component x is a real number, that each second component w is a complex number, and that no two pairs have the same first component. A *function* f, in Chapters 2 through 6, is a complex-valued function of a real variable, unless the context indicates otherwise. What has been said about functions in Subsection 1.1.2 carries over, with some obvious changes in the formulations, to the new situation. Hence, it is sufficient here to give a few remarks.

A complex-valued function f has the representation

$$f(x) = u(x) + iv(x), \tag{1}$$

where $u(x) = \operatorname{Re} f(x)$ and $v(x) = \operatorname{Im} f(x)$ are real-valued functions with D_f as domain. The function $f(x)$ is *integrable* on the interval (a, b) if $u(x)$ and $v(x)$ are both integrable on (a, b), and its *integral* on (a, b), denoted $\int_a^b f(x)\, dx$, is defined by

$$\int_a^b f(x)\, dx = \int_a^b u(x)\, dx + i \int_a^b v(x)\, dx. \tag{2}$$

Further, $f(x)$ is *square integrable* on (a, b) if $f(x)$ and $|f(x)|^2$ are integrable on (a, b).

EXERCISES

201. Evaluate the integrals (the identity $e^{ix} = \cos x + i \sin x$ may be useful)

a) $\displaystyle\int_0^1 (1 + ix)^2\, dx$,

b) $\displaystyle\int_{-1}^1 (3 - 2ix)^3\, dx$,

c) $\displaystyle\int_0^{\pi/2} e^{ix}\, dx$,

d) $\displaystyle\int_0^{2\pi} e^{-(2+i)x}\, dx$.

*202. Suppose that $P(x)$ is a polynomial with complex coefficients. Find a necessary

26

and sufficient condition for the function $1/P(x)$ to be integrable on every finite interval.

203. Suppose that $f(x)$ is a complex-valued function integrable on an interval (a, b). Show that the function $|f(x)|$ is integrable on (a, b), and that $\left|\int_a^b f(x)\, dx\right| \leqslant \int_a^b |f(x)|\, dx$. *Hint.* Let index m denote truncation by the lines $y = m$ and $y = -m$ as in the proof of Theorem 1.5.4. Show that, using the notation (1), for every x in (a, b)

$$\begin{aligned}
|f(x)|_m &\leqslant [(u_m(x))^2 + (v_m(x))^2]^{1/2} \\
&\leqslant |u_m(x)| + |v_m(x)| \leqslant |u(x)| + |v(x)|.
\end{aligned}$$

Prove that the left and right members are integrable on (a, b), and deduce that

$$\lim_{m \to \infty} \int_a^b |f(x)|_m\, dx \text{ exists. Then prove the assertions.}$$

204. Suppose that $f(x)$ and $g(x)$ are complex-valued functions, square integrable on an interval (a, b). Show that their product $f(x)g(x)$ is integrable on the interval. *Hint.* Use the notation (1) and, analogously, set $g(x) = u_1(x) + iv_1(x)$. Then $fg = uu_1 - vv_1 + i(uv_1 + u_1v)$. Apply the result in Exercise 115 to each of the terms in the right member.

205. Suppose that $f(x)$ is a complex-valued function, square integrable on the interval $(-\pi, \pi)$ and having the Fourier coefficients a_n and b_n. Prove the following variant of the Parseval formula in Theorem 1.5.4

$$\tfrac{1}{2}|a_0|^2 + \sum_{n=1}^{\infty} (|a_n|^2 + |b_n|^2) = \frac{1}{\pi} \int_{-\pi}^{\pi} |f(x)|^2\, dx.$$

2.2 ORTHOGONAL SYSTEMS

2.2.1 Introductory remark

It is clear from the discussion in Chapter 1 that the set of functions

$$\tfrac{1}{2}, \cos x, \sin x, \cos 2x, \sin 2x, \ldots, \cos kx, \sin kx, \ldots \tag{1}$$

is of fundamental importance in the theory of Fourier series. We list some properties of the functions (1):

a) Each function of the set (1) is bounded and integrable on the interval $(-\pi, \pi)$.

b) Each function $\phi_n(x)$ of the set (1) satisfies the condition $\int_{-\pi}^{\pi} \phi_n^2(x)\, dx > 0$.

c) Any two functions $\phi_m(x)$ and $\phi_n(x)$ of the set (1) satisfy, by (5) in Section 1.2.4, the equation $\int_{-\pi}^{\pi} \phi_m(x)\phi_n(x)\, dx = 0$.

Similar properties are shared by some other important sets of functions. This chapter is devoted to a study of such sets.

2.2.2 Definitions

We start this study by introducing some convenient terms and notations. We suppose that all functions considered have the interval (a, b) as domain.

Suppose that the complex-valued function $f(x)$ is square integrable on (a, b). The *norm* of $f(x)$, denoted by $\| f \|$, is defined by

$$\| f \| = \left[\int_a^b |f(x)|^2 dx \right]^{1/2}. \tag{2}$$

This definition shows that the norm is a nonnegative real number.

Suppose that the complex-valued functions $f(x)$ and $g(x)$ are integrable on (a, b). Also suppose that the product $f(x)\overline{g(x)}$ is integrable on the interval (this is for example the case if either $f(x)$ or $g(x)$ is bounded, or if both are square integrable; cf. Exercise 204). The *scalar product* of $f(x)$ and $g(x)$, denoted by (f, g), is defined by

$$(f, g) = \int_a^b f(x)\overline{g(x)}\, dx. \tag{3}$$

It is seen that

$$\| f \|^2 = (f, f). \tag{4}$$

We number the functions of each set of functions to be discussed by an index set, denoted by I. Here I is one of the three sets Z^+, N, Z, mentioned in Section 1.2.4.

Consider a system

$$\phi_n(x), \qquad n \in I, \; x \in (a, b), \tag{5}$$

consisting of a set of complex-valued functions $\phi_n(x)$, $n \in I$, and of an interval (a, b). The system (5) is called an *orthogonal system*, if it has the following three properties (similar to the properties in Remark 2.2.1):

1. Each function of the system is square integrable on (a, b).
2. Each function of the system has a positive norm.
3. Any two functions of the system are *orthogonal*, i.e. they have the scalar product zero.

If, further, each function of the system (5) has the norm 1, then (5) is called an *orthonormal system*. A given orthogonal system (5) can obviously be normalized into an orthonormal system by replacing the function $\phi_n(x)$ by the function $\phi_n(x)/\| \phi_n \|$ for each $n \in I$.

The absolute value sign in formula (2) and the bar in formula (3), denoting complex conjugation, are introduced because complex-valued functions are considered.   If the functions in (5) are real-valued, the absolute value sign and the bar can be deleted.

The adjective "orthogonal" should remind the reader of a property of vectors in R^3 (the set of ordered triples (a_1, a_2, a_3) of real numbers): Two vectors $\mathbf{a} = (a_1, a_2, a_3)$ and $\mathbf{b} = (b_1, b_2, b_3)$ are orthogonal (perpendicular) if and only if their scalar product $a_1 b_1 + a_2 b_2 + a_3 b_3$ is equal to zero; here the sum is somewhat similar to the integral in (3).

The reader should also observe that the norm has properties similar to those of the distance between points in R^3 (see Exercise 231).

2.2.3 Definitions

Suppose that (5) is an orthogonal system. Also suppose that the functions in (5) are continuous on the closed interval $[a, b]$. Then the arguments leading up to the formulas (6) in Definition 1.2.5 can be copied (Exercise 216). This leads us to introduce the following terminology.

Suppose that (5) is an orthogonal system, $f(x)$ is a function integrable on (a, b), and the products $f(x) \overline{\phi_n(x)}$, $n \in I$, are integrable on (a, b). Then the numbers

$$c_n = \frac{1}{\|\phi_n\|^2} (f, \phi_n), \qquad n \in I, \tag{6}$$

are called the *Fourier coefficients* of $f(x)$ with respect to the system (5). If the system is orthonormal, the formula (6) is simplified to

$$c_n = (f, \phi_n), \qquad n \in I. \tag{6'}$$

The series $\sum_{n \in I} c_n \phi_n(x)$ with the numbers (6) as coefficients is called the *Fourier series* of $f(x)$ with respect to the system (5). To indicate that $f(x)$ has this series as its Fourier series, the notation

$$f(x) \sim \sum_{n \in I} c_n \phi_n(x) \tag{7}$$

is used. The formation of the sum (7) for a given function $f(x)$ is called the *expansion* of $f(x)$ in a Fourier series with respect to the system (5). The nth *partial sum* $s_n(x)$ of the series in (7) is defined by

$$s_n(x) = \sum_{v=1}^{n} c_v \phi_v(x), \qquad s_n(x) = \sum_{v=0}^{n} c_v \phi_v(x), \tag{8a, b}$$

$$\text{or} \quad s_n(x) = \sum_{v=-n}^{n} c_v \phi_v(x), \tag{8c}$$

according as $I = Z^+$, $I = N$, or $I = Z$ respectively. The Fourier series in (7) is said to be *convergent* at a given point x, if the limit $\lim_{n \to \infty} s_n(x)$ exists, and this limit is called the *sum* of the series at the point x and is denoted by the right member of (7).

In this definition we demand in the case $I = Z$, for practical reasons, that the *symmetric* partial sum converge as $n \to \infty$. Examples can be constructed where (8c) has a limit while $\sum_{v=-m}^{n} c_v \phi_v(x)$ has no limit as m and $n \to \infty$ independently. In our applications it will always be true that $c_n \phi_n(x) \to 0$ as $|n| \to \infty$; hence, such unacceptable equations as $\sum_{n \in Z} n^3 = 0$ will not occur. The limit of the symmetric partial sum can be denoted by pr.v. $\sum_{n \in Z} c_n \phi_n(x)$, where pr.v. stands for "the principal value of". We shall not use this more precise terminology.

We shall study some examples of orthogonal systems. The proofs of the assertions in the examples are left as exercises for the reader.

2.2.4 Example

The set (1) and the interval $(-\pi, \pi)$ form an orthogonal system. It is called the *trigonometric system.* The squares of the norms are

$$\left\|\tfrac{1}{2}\right\|^2 = \frac{\pi}{2}, \quad \|\cos nx\|^2 = \|\sin nx\|^2 = \pi \quad \text{for } n \in Z^+.$$

Formulas (6) and (7) above give the formulas (6) and (7) of Definition 1.2.5.

2.2.5 Example

The system

$$\tfrac{1}{2}, \cos x, \cos 2x, \ldots, \cos nx, \ldots, x \in (0, \pi), \tag{9}$$

is an orthogonal system, called the *cosine system.* The corresponding Fourier coefficients of a function $f(x)$, integrable on the interval $(0, \pi)$, are denoted by a_n:

$$a_n = \frac{2}{\pi} \int_0^\pi f(x) \cos nx \, dx, \qquad n \in N. \tag{10}$$

The formula (7) is here

$$f(x) \sim \tfrac{1}{2}a_0 + \sum_{n=1}^{\infty} a_n \cos nx. \tag{11}$$

The numbers (10) are called the *cosine coefficients* of the function, and the series in (11) is called its *cosine series.*

2.2.6 Example

The system

$$\sin x, \sin 2x, \ldots, \sin nx, \ldots, x \in (0, \pi), \tag{12}$$

is an orthogonal system, called the *sine system.* The corresponding Fourier coefficients of a function $f(x)$, integrable on the interval $(0, \pi)$, are denoted by b_n:

$$b_n = \frac{2}{\pi} \int_0^\pi f(x) \sin nx \, dx, \qquad n \in Z^+. \tag{13}$$

The formula (7) is here

$$f(x) \sim \sum_{n=1}^{\infty} b_n \sin nx. \tag{14}$$

The numbers (13) are called the *sine coefficients* of the function, and the series in (14) is called its *sine series.*

2.2.7 Example

The system

$$\ldots, e^{-2ix}, e^{-ix}, 1, e^{ix}, e^{2ix}, \ldots, e^{inx}, \ldots, x \in (-\pi, \pi), \tag{15}$$

is an orthogonal system, called the *complex trigonometric system.* For brevity we shall call it the *complex system.* The corresponding Fourier coefficients of a function $f(x)$, integrable on the interval $(-\pi, \pi)$, are denoted by c_n:

$$c_n = \frac{1}{2\pi} \int_{-\pi}^{\pi} f(x)e^{-inx}\, dx, \qquad n \in Z. \tag{16}$$

The formula (7) is here

$$f(x) \sim \sum_{n=-\infty}^{\infty} c_n e^{inx}. \tag{17}$$

The numbers (16) are called the *complex Fourier coefficients* of the function, and the series in (17) is called its *complex Fourier series.* The complex Fourier coefficients of $f(x)$ can be expressed in terms of its ordinary Fourier coefficients (defined by (6) in Definition 1.2.5) and vice versa by means of the formulas (for $n = 0$ in (18a), set $b_n = 0$)

$$c_n = \tfrac{1}{2}(a_n - ib_n) \qquad \text{for} \quad n \geqslant 0, \tag{18a}$$

$$c_{-n} = \tfrac{1}{2}(a_n + ib_n) \qquad \text{for} \quad n > 0, \tag{18b}$$

$$a_n = c_n + c_{-n} \qquad \text{for} \quad n \geqslant 0, \tag{19a}$$

$$b_n = i(c_n - c_{-n}) \qquad \text{for} \quad n > 0. \tag{19b}$$

2.2.8 Example

Suppose that a is a positive number. The system

$$\tfrac{1}{2}, \cos \frac{\pi x}{a}, \sin \frac{\pi x}{a}, \ldots, \cos \frac{n\pi x}{a}, \sin \frac{n\pi x}{a}, \ldots, x \in (-a, a), \tag{20}$$

is an orthogonal system. The corresponding Fourier coefficients of a function $f(x)$, integrable on the interval $(-a, a)$, are denoted by a_n and b_n:

$$a_n = \frac{1}{a} \int_{-a}^{a} f(x) \cos \frac{n\pi x}{a}\, dx, \qquad n \in N, \tag{21a}$$

$$b_n = \frac{1}{a} \int_{-a}^{a} f(x) \sin \frac{n\pi x}{a}\, dx, \qquad n \in Z^{+}. \tag{21b}$$

The formula (7) is here

$$f(x) \sim \tfrac{1}{2}a_0 + \sum_{n=1}^{\infty} \left(a_n \cos \frac{n\pi x}{a} + b_n \sin \frac{n\pi x}{a} \right). \tag{22}$$

EXERCISES

206. a) Show that the cosine system (9) is an orthogonal system.
      b) Prove formula (10).
      c) State and prove a convergence theorem for the system (9), analogous to Theorem 1.4.4.

207. a) Show that the sine system (12) is an orthogonal system.
      b) Prove formula (13).
      c) State and prove a convergence theorem for the system (12), analogous to Corollary 1.4.5.
      d) Normalize the system (9) into an orthonormal system.

208. a) Expand the function $f(x) = e^x$, $0 < x < \pi$, in a cosine series.
      b) Draw a figure, showing the sum of the cosine series on the interval $[-2\pi, 2\pi]$.
      c) Do the same things for the sine series of the function.

209. a) Show that the complex system (15) is an orthogonal system.
      b) Prove formula (16).
      c) Prove formulas (18) and (19).
      d) Suppose that the function $f(x)$ is integrable on the interval $(-\pi, \pi)$, and let $s_n(x)$ and $t_n(x)$ denote the nth partial sum of its Fourier series with respect to the systems (1) and (15) respectively. Show that $s_n(x) = t_n(x)$ for each n and each x.
      e) State and prove a convergence theorem for the system (15), analogous to Theorem 1.4.4.

210. Expand the function $f(x) = \pi - x$, $0 < x < 2\pi$, in a complex Fourier series.

211. Suppose that $f(x)$ is a real-valued function, integrable on the interval $(-\pi, \pi)$. Show that its complex Fourier coefficients satisfy the equation $c_{-n} = \overline{c_n}$ for each n.

212. a) Show that the system (20) is an orthogonal system.
      Hint. Introduce a new independent variable t by the substitution $at = \pi x$.
      b) Prove the formulas (21).
      c) State and prove a convergence theorem for the system (20), analogous to Theorem 1.4.4.

213. Expand the function $f(x) = 4x - x^2$, $0 < x < 4$, in a Fourier series that only contains (a) sine terms, (b) cosine terms. Draw in each case a figure that shows the sum of the series on the interval $[-8, 8]$.

214. Suppose that $a > 0$. Expand the function $|\sin \pi x/a|$ in a Fourier series with respect to the system (20).

215. a) Show that the system
$$\sin(2n - 1)x, \qquad n \in Z^+. \ x \in (0, \pi/2),$$
      is an orthogonal system.
      b) Deduce a formula for the corresponding Fourier coefficients b_{2n-1}, $n \in Z^+$.

*216. Give the arguments indicated in the first paragraph of Definition 2.2.3.

217. Suppose that the complex-valued function $f(x)$ is square integrable on the interval $(0, \pi)$ and that it has the cosine coefficients a_n and the sine coefficients b_n. Show that $\frac{1}{2}|a_0|^2 + \sum_{n=1}^{\infty}|a_n|^2 = \sum_{n=1}^{\infty}|b_n|^2$.
      Hint. Apply Exercise 205 to functions, conveniently defined on $(-\pi, \pi)$.

218. Suppose that $f_0(x)$ is an odd function of period 2, that $f_0(x) = 1$ for $0 < x < 1$, and that no integer belongs to the domain of $f_0(x)$. Suppose that $f_n(x) = f_0(2^n x)$ for $n \in Z^+$.
   a) Draw figures showing the functions $f_0(x), f_1(x), f_2(x), f_3(x)$ on the interval $(0, 1)$.
   b) Show that the system $f_n(x)$, $n \in N$, $x \in (0, 1)$, is orthonormal. (It is called the Rademacher system.)

2.3 COMPLETE ORTHOGONAL SYSTEMS

Theorem 2.3.1 below is a generalization of Problem 1.3.3 and its solution to general orthogonal systems. Similarly, Theorem 2.3.3 is an extension of the inequality (4) of Remark 1.3.4. The formula obtained is also called the Bessel inequality. We shall use the notation "$|v| \leqslant n$" to mean "$v \in I$ and $|v| \leqslant n$".

   2.3.1 Theorem (showing a minimizing property of the Fourier coefficients). *Suppose that $\phi_m(x)$, $m \in I$, $x \in (a, b)$, is an orthogonal system, that $f(x)$ is a complex-valued function, square integrable on the interval (a, b), that $f(x) \sim \sum_{m \in I} c_m \phi_m(x)$, that n is a given nonnegative integer in I, and that $d_v, |v| \leqslant n$, are given complex numbers. Then*

$$\left\| f(x) - \sum_{|v| \leqslant n} c_v \phi_v(x) \right\| \leqslant \left\| f(x) - \sum_{|v| \leqslant n} d_v \phi_v(x) \right\| \tag{1}$$

with equality if and only if $d_v = c_v$ for each v.

   An alternative formulation of the assertion is as follows: the square deviation of $\sum_{|v| \leqslant n} d_v \phi_v(x)$ from $f(x)$ is minimal if and only if this sum is the nth partial sum of the Fourier series of $f(x)$.

   Proof (copying the solution of Problem 1.3.3). We have, writing $\sum$ for $\sum_{|v| \leqslant n}$,

$$\left\| f - \sum d_v \phi_v \right\|^2 = (f - \sum d_v \phi_v, f - \sum d_v \phi_v)$$
$$= (f, f) - \sum \overline{d_v}(f, \phi_v) - \sum d_v \overline{(f, \phi_v)} + \sum d_v \overline{d_v}(\phi_v, \phi_v)$$
$$= (f, f) - \sum c_v \overline{c_v} \| \phi_v \|^2 + \sum (c_v - d_v)(\overline{c_v} - \overline{d_v}) \| \phi_v \|^2,$$

and hence

(a) $\left\| f(x) - \sum_{|v| \leqslant n} d_v \phi_v(x) \right\|^2$

$$= \| f(x) \|^2 - \sum_{|v| \leqslant n} |c_v|^2 \| \phi_v \|^2 + \sum_{|v| \leqslant n} |c_v - d_v|^2 \| \phi_v \|^2.$$

The equation (a) gives the assertion.

   2.3.2 Corollary. *Under the hypotheses of Theorem 2.3.1 and defining $s_n(x)$ by Definition 2.2.3, we have*

$$\| f(x) - s_n(x) \|^2 = \| f(x) \|^2 - \sum_{|v| \leqslant n} |c_v|^2 \| \phi_v \|^2. \tag{2}$$

Proof. Replace d_v by c_v in (a) for each v. Then (2) is obtained.

2.3.3 Theorem (giving the Bessel inequality). *Under the hypotheses of Theorem 2.3.1 there holds*

$$\sum_{n\in I} |c_n|^2 \|\phi_n\|^2 \leqslant \|f(x)\|^2. \tag{3}$$

Proof. The left member of (2) is nonnegative for each n. Let $n \to \infty$ in the right member. Then (3) follows.

Can the inequality sign in the Bessel inequality (3) be replaced by an equality sign? The following example shows that this need not be the case. In Theorem 2.3.3, set $I = Z^+$, $\phi_n(x) = \cos nx$, $(a, b) = (0, \pi)$, and $f(x) = 1$. Then the left member of (3) is zero, and the right member is π. This observation leads us to introduce two concepts—the Parseval property and complete orthogonal system—and to give a theorem.

2.3.4 Definitions

We shall say that an orthogonal system $\phi_n(x)$, $n \in I$, $x \in (a, b)$, has the *Parseval property*, if for any complex-valued function $f(x)$, square integrable on (a, b), its Fourier coefficients c_n with respect to the system satisfy the equation

$$\sum_{n\in I} |c_n|^2 \|\phi_n\|^2 = \|f(x)\|^2. \tag{4}$$

This equation is then called the *Parseval formula*.

An orthogonal system is said to be *complete*, if no complex-valued function, square integrable on the interval of the system, can be added to the system with preservation of Properties 1, 2, 3 in Definition 2.2.2.

2.3.5 Theorem. *If an orthogonal system has the Parseval property, then it is complete.*

Proof. Suppose that $\phi_n(x)$, $n \in I$, $x \in (a, b)$, is a given orthogonal system that has the Parseval property. Add to the system a function $\phi(x)$ such that Properties 1 and 3 of Definition 2.2.2 hold in the enlarged system. Let $\sum_{n\in I} c_n\phi_n(x)$ be the Fourier series of $\phi(x)$ with respect to the given system. By Property 3, every c_n is equal to zero. Formula (4) then shows that $\|\phi(x)\| = 0$. This contradicts Property 2. Hence the assertion holds.

The converse of this theorem holds for the class of Lebesgue-integrable functions (see [14], p. 22, Theorem 22).

Theorems 2.3.5 and 1.5.4 give a corollary for the system of Example 2.2.4.

2.3.6 Corollary. *The trigonometric system* $\frac{1}{2}$, $\cos x$, $\sin x$, . . . , $\cos nx$, $\sin nx$, . . . , $x \in (-\pi, \pi)$, *is complete.*

Are the systems of Examples 2.2.5–2.2.8 also complete? We shall study the complex system in an example and a corollary. For the remaining systems the questions of completeness are left as exercises.

2.3.7 Example

Suppose that $f(x)$ is a complex-valued function, square integrable on the interval

$(-\pi, \pi)$. Denote by $\sum_{n \in Z} c_n e^{inx}$ the complex Fourier series of $f(x)$. Prove the Parseval formula

$$\sum_{n=-\infty}^{\infty} |c_n|^2 = \frac{1}{2\pi} \int_{-\pi}^{\pi} |f(x)|^2 \, dx. \tag{5}$$

Solution. By (18) in Example 2.2.7 and the parallelogram law (the extension of the hint in Exercise 122 to complex numbers), the left member of (5) is equal to

$$\sum_{n=1}^{\infty} \tfrac{1}{4}|a_n + ib_n|^2 + \tfrac{1}{4}|a_0|^2 + \sum_{n=1}^{\infty} \tfrac{1}{4}|a_n - ib_n|^2 = \tfrac{1}{4}|a_0|^2 + \tfrac{1}{2}\sum_{n=1}^{\infty}(|a_n|^2 + |b_n|^2).$$

Now Exercise 205 gives the assertion.

2.3.8 Corollary. *The complex system* $\ldots, e^{-ix}, 1, e^{ix}, \ldots, e^{inx}, \ldots, x \in$ $(-\pi, \pi)$, *is complete.*

Proof. Apply Example 2.3.7 and Theorem 2.3.5.

We give a definition and a theorem that generalizes the result of Exercise 127.

2.3.9 Definition

Consider functions, square integrable on the interval (a, b). A sequence of functions $s_n(x)$, $n \in N$ or $n \in Z^+$, is said to *converge in the mean* to a function $f(x)$ if $\| f(x) - s_n(x) \| \to 0$ as $n \to \infty$.

2.3.10 Theorem. *Given an orthogonal system having the Parseval property and given a function $f(x)$, square integrable on the interval of the system, the sequence of partial sums $s_n(x)$ of the Fourier series of $f(x)$ with respect to the system converges in the mean to $f(x)$.*

Proof. The assertion follows from Corollary 2.3.2.

EXERCISES

219. Is the cosine system (see Example 2.2.5) complete?
  Hint. Let $f(x)$ be a function, square integrable on $(0, \pi)$. Extend $f(x)$ to an even function. Apply Exercise 205 and Theorem 2.3.5.

220. Is the sine system (see Example 2.2.6) complete?

221. Is the system of Example 2.2.8 complete?

*222. Define $f(x)$ and $s_n(x)$ on the interval $(0, \pi)$ as in Fig. 1.5. By Theorem 2.3.10, $\| f(x) - s_n(x) \| \to 0$ as $n \to \infty$. Find the smallest n for which $\| f(x) - s_n(x) \| < \tfrac{1}{2}$.

223. Suppose that a is a real nonintegral number. Expand the function $f(x) = e^{iax}$, $|x| < \pi$, in a complex Fourier series. Use the result and the Parseval formula (5) to prove that

$$\sum_{n=-\infty}^{\infty} \frac{1}{(n-a)^2} = \frac{\pi^2}{\sin^2 a\pi}.$$

224. Suppose that a is a positive number. Prove the equality

$$\sum_{n=1}^{\infty} \frac{1}{n^2 + a^2} = \frac{\pi}{2a} \coth a\pi - \frac{1}{2a^2}$$

by expanding the function $f(x) = e^{ax}$, $0 < x < 2\pi$, in a complex Fourier series and then applying the Parseval formula.

225. Is the system of Exercise 215 complete?
    Hint. Suppose that the function $f(x)$ is square integrable on $(0, \pi/2)$. Extend $f(x)$ to an odd function such that the function $g(x)$ defined by $g(x) = f(x + \pi/2)$ is an even function. Apply Exercise 205 and Theorem 2.3.5.

226. Is the Rademacher system (see Exercise 218) complete?

2.4 INTEGRATION OF FOURIER SERIES

We shall give a theorem on term-by-term integration of a Fourier series on an arbitrary interval. To this end we need two lemmas. The inequality (1) in Lemma 2.4.1 is called the Schwarz inequality. Exercise 204 shows that the left member of (1) is defined. Equation (2) in Lemma 2.4.2 is obviously an extension of the Parseval formula in Definition 2.3.4. It is known that, for the orthogonal systems in Examples 2.2.4–2.2.8, the conclusion of Theorem 2.4.3 holds true if $f(x)$ is integrable on (a, b), not necessarily square integrable (see [19]); in this text we shall not prove this property.

2.4.1 Lemma (giving the Schwarz inequality). *Suppose that $f(x)$ and $g(x)$ are complex-valued functions, square integrable on an interval. Then*

$$|(f, g)| \leqslant \| f \| \| g \|. \tag{1}$$

Proof. Set $\theta = \arg(f, g)$. Then $\mathrm{Re}(fe^{-i\theta}, g) = |(f, g)|$. Let λ denote a real number. Then

$$\begin{aligned}
0 \leqslant \| fe^{-i\theta} + \lambda g \|^2 &= (fe^{-i\theta} + \lambda g, fe^{-i\theta} + \lambda g)\\
&= (fe^{-i\theta}, fe^{-i\theta}) + \lambda(fe^{-i\theta}, g) + \lambda\overline{(fe^{-i\theta}, g)} + \lambda^2(g, g)\\
&= \| fe^{-i\theta} \|^2 + 2\lambda \mathrm{Re}(fe^{-i\theta}, g) + \lambda^2 \| g \|^2\\
&= \| f \|^2 + 2\lambda |(f, g)| + \lambda^2 \| g \|^2.
\end{aligned}$$

First suppose that $\| g \| > 0$. Then the right member of (a) is a real-valued polynomial of degree 2 in λ. The range of this polynomial contains no negative number. Hence its discriminant is negative or zero: $|(f, g)|^2 - \| f \|^2 \| g \|^2 \leqslant 0$. This proves the assertion for $\| g \| > 0$. Then suppose that $\| g \| = 0$. As (a) holds for every real number λ, $(f, g) = 0$. Hence the assertion is true also in this case.

2.4.2 Lemma (giving an extension of the Parseval formula). *Suppose that $\phi_n(x)$, $n \in I$, $x \in (a, b)$, is an orthogonal system having the Parseval property. Suppose that $f(x)$ and $g(x)$ are complex-valued functions, square integrable on (a, b). Denote their Fourier coefficients with respect to the system by c_n and d_n respectively. Then*

$$(f, g) = \sum_{n \in I} c_n \overline{d_n} \| \phi_n \|^2. \tag{2}$$

Proof. Denote by $s_n(x)$ the nth partial sum of the Fourier series of $f(x)$. Lemma 2.4.1 and Corollary 2.3.2 give

$$|(f - s_n, g)| \leqslant \|f - s_n\| \|g\| \to 0 \qquad \text{as} \quad n \to \infty.$$

This property and

$$(f, g) = (f - s_n, g) + (s_n, g)$$

yield

$$(f, g) = \lim_{n \to \infty} (s_n, g) = \lim_{n \to \infty} \sum_{|v| \leqslant n} c_v(\phi_v, g) = \lim_{n \to \infty} \sum_{|v| \leqslant n} c_v \overline{d_v} \|\phi_v\|^2.$$

This is the assertion.

2.4.3 Theorem. *Under the assumptions of Lemma 2.4.2 there holds for any α, β such that $a \leqslant \alpha < \beta \leqslant b$*

$$\int_\alpha^\beta f(x)\, dx = \sum_{n \in I} c_n \int_\alpha^\beta \phi_n(x)\, dx. \tag{3}$$

Proof. In Lemma 2.4.2, set $g(x) = 1$ for $\alpha < x < \beta$ and $g(x) = 0$ otherwise. Then

$$\overline{d_n} \|\phi_n\|^2 = \int_a^b \overline{g(x)}\, \phi_n(x)\, dx = \int_\alpha^\beta \phi_n(x)\, dx.$$

Formula (2) now gives the assertion.

EXERCISES

227. Use the results of Examples 1.5.5 and 1.6.3 and Exercise 132 to compute $\sum_{n=1}^\infty n^{-2} \cos nx$ and $\sum_{n=1}^\infty n^{-2} \sin nx$ for $0 \leqslant x \leqslant 2\pi$.

228. Express the integral $\int_{-\pi}^\pi |x| \cosh x\, dx$ in terms of the ordinary Fourier coefficients (i.e. Fourier coefficients with respect to the trigonometric system) of the functions $f(x) = |x|$, $|x| < \pi$, and $g(x) = \cosh x$, $|x| < \pi$.

*229. Suppose that $f(x)$ and $g(x)$ are complex-valued functions, continuous on the closed interval $[a, b]$. Find a necessary and sufficient condition for an equality sign to hold in the Schwarz inequality.

230. Suppose that $f(x)$ is a complex-valued function, square integrable on the interval (a, b). Show that

$$\left[\int_a^b |f(x)|\, dx\right]^2 \leqslant (b - a)\int_a^b |f(x)|^2\, dx.$$

231. Suppose that $f(x)$ and $g(x)$ are complex-valued functions, square integrable on the interval (a, b), and that c is a complex number. Show that
a) $f(x) = 0$ for $a < x < b \Rightarrow \|f\| = 0$,
b) $\|cf\| = |c| \|f\|$,
c) $\|f + g\| \leqslant \|f\| + \|g\|$, the *triangle inequality*.
Hint for (c). Observe that $\|f + g\|^2 = (f + g, f + g) = (f, f) + (f, g) + (g, f) + (g, g) = \|f\|^2 + 2\,\mathrm{Re}\,(f, g) + \|g\|^2$. Apply the Schwarz inequality.

2.5 THE GRAM-SCHMIDT ORTHOGONALIZATION PROCESS

In this section (2.5) we restrict our discussion to functions continuous on a closed interval. Let a sequence of functions $f_n(x)$, $n \in N$, be given, satisfying a certain natural condition. We shall show how to construct functions $\phi_0(x)$, $\phi_1(x)$, ... of an orthogonal system, such that each function $\phi_n(x)$ is expressed in a simple way in terms of the first $n + 1$ functions of the given sequence. The construction is called the Gram-Schmidt orthogonalization process.

2.5.1 Definitions

Suppose that all functions considered are complex-valued and continuous, and that they have the closed interval $[a, b]$ as domain.

A finite sequence of functions $f_0(x), f_1(x), \ldots, f_n(x)$, where $n \geqslant 0$, is said to be *linearly dependent* if there exist complex numbers c_v, not all zero, such that $\sum_{v=0}^{n} c_v f_v(x) = 0$ for $a \leqslant x \leqslant b$. A sequence of functions $f_n(x)$, $n \in N$, is said to be *linearly dependent* if it has a linearly dependent finite subsequence. A finite or infinite sequence of functions is said to be *linearly independent*, if it is not linearly dependent.

A function $\phi(x)$ is called a *linear combination* of the functions $f_0(x), f_1(x), \ldots, f_n(x)$, where $n \geqslant 0$, if there exist complex numbers c_v such that $\phi(x) = \sum_{v=0}^{n} c_v f_v(x)$ for $a \leqslant x \leqslant b$.

2.5.2 Theorem (giving the Gram-Schmidt orthogonalization process). *Suppose that $f_n(x)$, $n \in N$, $x \in [a, b]$, is a linearly independent sequence of continuous functions. Then there exists an orthogonal system $\phi_n(x)$, $n \in N$, $x \in [a, b]$, such that each $\phi_n(x)$ is a linear combination of the functions $f_v(x)$, $0 \leqslant v \leqslant n$. The functions $\phi_n(x)$ are uniquely determined up to nonzero factors.*

Proof. 1. Set $\phi_0 = f_0$. Then the function ϕ_0 fulfills the requirements on ϕ_0 in Definition 2.2.2. Further, ϕ_0 and f_0 are linear combinations of each other.

2. Suppose that the functions $\phi_0, \phi_1, \ldots, \phi_{n-1}$, $n \geqslant 1$, fulfill the requirements in Definition 2.2.2, that each ϕ_v is a linear combination of $f_0, f_1, \ldots, f_v$, and that each f_v, $0 \leqslant v \leqslant n - 1$, is a linear combination of $\phi_0, \phi_1, \ldots, \phi_v$. Set

$$\phi_n(x) = f_n(x) - \sum_{v=0}^{n-1} (f_n, \phi_v) \phi_v(x) / \| \phi_v \|^2 \qquad \text{for} \quad a \leqslant x \leqslant b. \qquad (1)$$

It is readily verified that the functions $\phi_0, \phi_1, \ldots, \phi_n$ have Properties 1 and 3 of Definition 2.2.2, that ϕ_n is a linear combination of $f_0, f_1, \ldots, f_n$ and that f_n is a linear combination of $\phi_0, \phi_1, \ldots, \phi_n$. Assume for the moment that ϕ_n does not have Property 2 of Definition 2.2.2. Then $\| \phi_n \| = 0$, $\phi_n(x) = 0$ for $a \leqslant x \leqslant b$, and formula (1) shows that f_n is a linear combination of $f_0, f_1, \ldots, f_{n-1}$. This contradicts the assumption that the finite sequence $f_0, f_1, \ldots, f_n$ is linearly independent. Hence ϕ_n has also Property 2 of Definition 2.2.2.

3. Suppose that the construction in parts 1 and 2 is carried out for each $n \in N$. Then $\phi_n(x)$, $n \in N$, $x \in [a, b]$, is an orthogonal system, and each $\phi_n(x)$ is a linear combination of the functions $f_v(x)$, $0 \leqslant v \leqslant n$.

4. Suppose that ϕ_n and ψ_n, $n \in N$, are two sequences of functions that satisfy the assertion in Theorem 2.5.2, perhaps with the exception of the clause on uniqueness. Then there exists a constant a_0 such that $\psi_0 = a_0\phi_0$. Suppose that there exist constants $a_0, a_1, \ldots, a_{n-1}$, $n \geqslant 1$, such that $\psi_v = a_v\phi_v$ for $0 \leqslant v \leqslant n-1$. Observe that f_n is a linear combination of $\phi_0, \phi_1, \ldots, \phi_n$ (for if not, the finite sequence $\phi_0, \phi_1, \ldots, \phi_n$ is linearly dependent; this contradicts the result in Exercise 236) and also a linear combination of $\psi_0, \psi_1, \ldots, \psi_n$. Hence there are constants b_v, c_v such that

$$f_n = b_0\phi_0 + \ldots + b_{n-1}\phi_{n-1} + b_n\phi_n,$$
$$f_n = c_0\phi_0 + \ldots + c_{n-1}\phi_{n-1} + c_n\psi_n.$$

Multiply both members of each equality by $\overline{\phi}_v$, $0 \leqslant v \leqslant n-1$, and integrate from a to b. It follows that $b_v = c_v$ for $0 \leqslant v \leqslant n-1$. Then $b_n\phi_n = c_n\psi_n$. Here $b_n \neq 0$ and $c_n \neq 0$. The last assertion now follows by induction.

EXERCISES

*232. Determine for each of the following finite or infinite sequences of functions whether it is linearly independent or not:
   a) $f_n(x) = e^{nx}$, $n \in \{1, 2, 3\}$, $x \in [0, 1]$,
   b) $f_n(x) = \sin(x + n\pi/4)$, $n \in \{0, 1, 2\}$, $x \in [-\pi, \pi]$,
   c) $f_n(x) = x^n$, $n \in N$, $x \in [-1, 1]$,
   d) $f_n(x) = \sin nx$, $n \in N$, $x \in [0, \pi]$.

233. Obtain an orthogonal system $\phi_n(x)$, $n \in N$, $x \in [0, \pi]$, such that $\phi_n(0) = 1$ for each n, by applying the Gram-Schmidt process to the sequence $f_n(x) = \cos^n x$.
   Hint. Use the binomial expansion of $\cos^n x = 2^{-n}(e^{ix} + e^{-ix})^n$ to show that $\cos^n x$ is a linear combination of $1, \cos x, \cos 2x, \ldots, \cos nx$.

234. In the previous exercise, the system obtained was normalized by prescribing the value 1 for each of the functions at a certain point. Is this normalization always feasible in applications of Theorem 2.5.2?

235. Find the first few functions, say four of them, of an orthogonal system $\phi_n(x)$, $n \in N$, $x \in [-1, 1]$, obtained by applying the Gram-Schmidt process to the sequence $f_n(x) = x^n$; normalize by letting $\phi_n(x)$ have the leading term x^n.

236. Show that a sequence of functions from an orthogonal system is linearly independent.

2.6 STURM–LIOUVILLE PROBLEMS

In this section (2.6) we restrict our discussion to functions having a closed interval [a, b] as domain. The derivative of a differentiable function $f(x)$ is often denoted by $f'(x)$; analogously $f''(x)$ denotes the second derivative. We shall give a brief account of a class of problems called Sturm–Liouville problems.

2.6.1 Introductory problem

Find those real numbers λ for which the boundary-value problem

$$\frac{d^2y}{dx^2} + \lambda y = 0, \qquad x \in [0, \pi], \tag{1a}$$

$$y(0) = y(\pi) = 0, \tag{1b}$$

has nontrivial solutions. Also find the nontrivial solutions corresponding to each permissible λ.

In (1a), a differential equation is given on the interval $[0, \pi]$. A solution of (1a) is a function $y(x)$ with domain $[0, \pi]$ such that $y''(x) + \lambda y(x) = 0$ for $0 \leqslant x \leqslant \pi$. A solution of the problem (1), i.e. (1a) combined with (1b), is a solution of (1a) that satisfies (1b). The λ in (1a) is a parameter. For each λ, problem (1) has the trivial solution: $y(x) = 0$ for $0 \leqslant x \leqslant \pi$. We are looking for those real values of λ, called the eigenvalues for the problem (1), for which there exist solutions $y(x)$, such that $y(x) \neq 0$ for some x in $[0,\pi]$; such solutions are called eigenfunctions for the problem. The interval $[0, \pi]$, considered as a set on the x-axis, has the points 0 and π as its boundary. This accounts for the term "boundary-value problem".

Solution. **1.** Suppose that the number λ and the function $y(x)$ satisfy the conditions of problem (1). Suppose $\lambda < 0$. Then (cf. Lemma 5.4.2) there exist constants c_1 and c_2 such that

$$y(x) = c_1 e^{\sqrt{-\lambda}x} + c_2 e^{-\sqrt{-\lambda}x}, \qquad x \in [0, \pi],$$
$$c_1 + c_2 = c_1 e^{\sqrt{-\lambda}\pi} + c_2 e^{-\sqrt{-\lambda}\pi} = 0.$$

It follows that $c_1 = c_2 = 0$. This contradicts the assumption that $y(x)$ is nontrivial. Suppose instead $\lambda = 0$. Then there exist constants c_1 and c_2 such that

$$y(x) = c_1 + c_2 x, \qquad c_1 = c_1 + c_2 \pi = 0,$$

and we have again the contradiction $c_1 = c_2 = 0$. Hence $\lambda > 0$. Then there exist constants c_1 and c_2, not both zero, such that

$$y(x) = c_1 \cos \sqrt{\lambda}x + c_2 \sin \sqrt{\lambda}x,$$
$$c_1 = c_1 \cos \sqrt{\lambda}\pi + c_2 \sin \sqrt{\lambda}\pi = 0,$$
$$c_1 = 0, \qquad c_2 \neq 0, \qquad \sin \sqrt{\lambda}\pi = 0.$$

It follows that $\sqrt{\lambda} \in Z^+$. Set $n = \sqrt{\lambda}$ and $c_n = c_2$. Then $\lambda = n^2$, $n \in Z^+$, and $y(x) = c_n \sin nx$ where c_n is a nonzero constant.

2. Suppose that $n \in Z^+$ and that c_n is a nonzero constant. It can be verified that the problem (1) is satisfied by $\lambda = n^2$ and $y(x) = c_n \sin nx$.

3. The results in **1** and **2** show that the problem (1) has the eigenvalues 1, 4, 9, ..., n^2, ... and that each eigenvalue n^2 has the corresponding eigenfunctions $y(x) = c_n \sin nx$, where c_n is a nonzero parameter.

2.6.2 Definitions

Suppose that the real-valued function $p(x)$ has a continuous derivative on the interval $[a, b]$, that the real-valued functions $q(x)$ and $w(x)$ are continuous on $[a, b]$, that $p(x) > 0$ and $w(x) > 0$ for $a \leqslant x \leqslant b$, that λ is a real parameter, that a_0, a_1, b_0, b_1 are real constants, that a_0, a_1 are not both zero, that b_0, b_1 are not both zero, and that

(a) $a_0 a_1 \leqslant 0$ and $b_0 b_1 \geqslant 0$.

Then the boundary-value problem

$$\frac{d}{dx}\left(p(x)\frac{dy}{dx}\right) - q(x)y + \lambda w(x)y = 0, \qquad x \in [a, b], \tag{2a}$$

$$a_0 y(a) + a_1 y'(a) = 0, \qquad b_0 y(b) + b_1 y'(b) = 0, \tag{2b, c}$$

is called a *Sturm–Liouville problem*. In (2a) a differential equation of second order is given on $[a, b]$. The function $w(x)$ is called a *weight function*. The conditions (2b) and (2c) are called the *boundary conditions*. Each value of λ for which the problem (2) has nontrivial solutions is called an *eigenvalue* for the problem. Each nontrivial solution is called an *eigenfunction* for the problem. It is immediately verified that if $\phi(x)$ is an eigenfunction and if c is a nonzero constant, then $c\phi(x)$ is also an eigenfunction. To *solve* the problem (2) is to find all its eigenvalues and all the corresponding eigenfunctions. Problem 2.6.1 is a Sturm–Liouville problem which we have just solved.

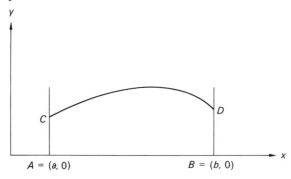

Figure 2.1

The definition of a Sturm–Liouville problem may seem complicated. The impetus to study such problems comes from physics. We sketch an example, using the notation of Fig. 2.1. Consider the problem of finding the motion of a vibrating string CD, whose density may be variable (say that the density at the point x is $w(x)$; here $w(x)$ is called the weight function) and whose end points C and D can slide without friction along the lines $x = a$ and $x = b$ respectively. Suppose that the end point C is attracted towards the point A by a force proportional to the distance AC; make an analogous assumption for the end point D. Suppose that, for each point x of the string, its deviation y from the equilibrium position on the x-axis and its velocity are known at the time $t = 0$. What is its deviation y at the time $t > 0$? By a (fairly realistic) mathematical idealization this problem can be formulated as a boundary-value problem for a certain partial differential equation. This problem can be reduced to a Sturm–Liouville problem. The forces acting at the point C have the resultant zero; this gives the first inequality (a). The second inequality (a) is explained analogously.

2.6.3 Remark

Suppose that the function $y(x)$ has a continuous second derivative. Set

$$L(y) = \frac{d}{dx}\left(p(x)\frac{dy}{dx}\right) - q(x)y. \tag{3}$$

Here L is an operator, i.e. a set of ordered pairs of functions $(y, L(y))$ such that no two pairs have the same first component. It can be verified (Exercise 237) that the operator L has the following property, called the Lagrange identity,

$$yL(z) - zL(y) = \frac{d}{dx}[p(x)(y(x)z'(x) - y'(x)z(x))]. \tag{4}$$

Integration yields the following formula, known as the Green formula,

$$\int_a^b [yL(z) - zL(y)]\,dx = \left[p(x)(y(x)z'(x) - y'(x)z(x))\right]_a^b. \tag{5}$$

2.6.4 Definitions
Consider a system

$$\phi_n(x),\ n \in I, \qquad w(x),\ x \in (a, b), \tag{6}$$

consisting of a set of functions $\phi_n(x)$, $n \in I$, of an interval (a, b), and of a function $w(x)$, such that $w(x) > 0$ for $a < x < b$; here $w(x)$ is called a weight function. Suppose for formula (7) below that the function $f(x)\sqrt{w(x)}$ is square integrable on (a, b), and for formula (8) below that the function $f(x)\overline{g(x)}w(x)$ is integrable on (a, b). Replace the formulas for the *norm* $\|f\|$ and the *scalar product* (f, g) in Definition 2.2.2 by

$$\|f\| = \left[\int_a^b |f(x)|^2 w(x)\,dx\right]^{1/2}, \tag{7}$$

$$(f, g) = \int_a^b f(x)\,\overline{g(x)}\,w(x)\,dx. \tag{8}$$

(Alternatively, consider (2) and (3) of Definition 2.2.2 as preliminary formulas, valid for the weight function $w(x) = 1$, $a < x < b$.) If the system (6) has the properties enumerated in Definition 2.2.2, then (6) is called an *orthogonal system with a weight function* $w(x)$. The rest of Definition 2.2.2 and Definition 2.2.3 can now be repeated with a few obvious changes in the formulations. In particular, if $f(x)$ is a given function such that the products $f(x)\overline{\phi_n(x)}w(x)$, $n \in I$, are integrable on (a, b), then $f(x)$ has *Fourier coefficients* c_n with respect to the system (6):

$$c_n = \frac{1}{\|\phi_n\|^2}\int_a^b f(x)\,\overline{\phi_n(x)}\,w(x)\,dx = \frac{1}{\|\phi_n\|^2}(f, \phi_n), \qquad n \in I, \tag{9}$$

where

$$\|\phi_n\|^2 = \int_a^b |\phi_n(x)|^2 w(x)\,dx, \qquad n \in I, \tag{10}$$

and a *Fourier series* with respect to the system (6):

$$f(x) \sim \sum_{n \in I} c_n \phi_n(x). \tag{11}$$

2.6.5 Theorem. *Suppose that the eigenvalues for a Sturm–Liouville problem* (2) *are* λ_m, $n \in Z^+$. *For each* n, *let* $\phi_n(x)$ *denote an eigenfunction corresponding to* λ_n. *Then:*
1. *The system* $\phi_n(x)$, $n \in Z^+$, $w(x)$, $x \in (a, b)$, *is an orthogonal system with the weight function* $w(x)$.
2. *For each* n, $\phi_n(x)$ *is uniquely determined up to a nonzero factor.*
3. *The set of eigenvalues* λ_m, $n \in Z^+$, *is bounded below.*

Proof. **1.** Suppose that λ_m and λ_n are two eigenvalues. By (2a) and (3)

$$L(\phi_m) + \lambda_m w(x)\phi_m(x) = 0, \quad L(\phi_n) + \lambda_n w(x)\phi_n(x) = 0,$$

whence

(b) $$\int_a^b [\phi_m L(\phi_n) - \phi_n L(\phi_m)]\, dx = (\lambda_m - \lambda_n) \int_a^b \phi_m(x)\phi_n(x)w(x)\, dx.$$

The Green formula (5) shows that the left member of (b) is equal to

(c) $$\left[p(x)\, (\phi_m(x)\phi_n'(x) - \phi_n(x)\phi_m'(x)) \right]_a^b.$$

By the boundary condition (2b), the vectors $(\phi_m(a), \phi_m'(a))$ and $(\phi_n(a), \phi_n'(a))$ are both perpendicular to the vector (a_0, a_1). Hence they are parallel. Substitution of the lower bound a into (c) then gives the value zero; similarly for the upper bound b. Hence the right member of (b) is zero. By assumption, $\lambda_m \neq \lambda_n$. Hence the integral in the right member of (b) is zero. This gives the first assertion.

2. Suppose that $\phi_n(x)$ and $\psi_n(x)$ are two eigenfunctions corresponding to the same eigenvalue λ_n. As in part **1** it is seen that the vectors $(\phi_n(a), \phi_n'(a))$ and $(\psi_n(a), \psi_n'(a))$ are parallel. Then there exist two constants c_1 and c_2, not both zero, such that

$$c_1 \phi_n(a) + c_2 \psi_n(a) = 0, \qquad c_1 \phi_n'(a) + c_2 \psi_n'(a) = 0.$$

Hence the function $c_1 \phi_n(x) + c_2 \psi_n(x)$ is a solution of (2a) for $\lambda = \lambda_n$ that vanishes together with its first derivative for $x = a$. Then (see Lemma 5.4.2) $c_1 \phi_n(x) + c_2 \psi_n(x) = 0$ for $x \in [a, b]$. This gives the second assertion.

3. Suppose that λ_n is an eigenvalue and $\phi_n(x)$ a corresponding real-valued eigenfunction. Then (2a) and integration by parts give

(d) $$\int_a^b \lambda_n \phi_n^2(x)w(x)\,dx = \int_a^b \phi_n(x) \left[q(x)\, \phi_n(x) - \frac{d}{dx}\, (p(x)\phi_n'(x)) \right] dx$$

$$= \int_a^b \left[q(x)\phi_n^2(x) + p(x)\phi_n'^2(x) \right] dx - p(x) \left[\phi_n(x)\phi_n'(x) \right]_a^b.$$

The conditions (a) and (2b, c) show that

$$\phi_n(a)\phi_n'(a) \geqslant 0 \qquad \text{and} \qquad -\phi_n(b)\phi_n'(b) \geqslant 0.$$

Then, by (d),

$$\lambda_n \int_a^b \phi_n^2(x)w(x)\, dx \geqslant \int_a^b q(x)\phi_n^2(x)\, dx.$$

It follows that $\lambda_n \geqslant [\min q(x)]/[\max w(x)]$, where $\min q(x)$ and $\max w(x)$ are for $a \leqslant x \leqslant b$. This proves the third assertion.

The theorem just proved can be extended to the following theorem, which we do not prove. For methods of proof see [16], Chapter 1, and [18], Chapter 2.

Theorem 2.6.6 deals with convergence in the mean. There are also theorems on pointwise convergence (cf. [16], Theorem 1.9).

2.6.6 Theorem. *Suppose that a Sturm–Liouville problem* (2) *is given. Then there exists a countably infinite set of eigenvalues for the problem. There exists an arrangement of the eigenvalues in an increasing sequence λ_n, $n \in Z^+$ (i.e. $\lambda_n < \lambda_{n+1}$ for each n). Further $\lambda_n \to \infty$ as $n \to \infty$. For each n, let $\phi_n(x)$ be an eigenfunction, corresponding to λ_n. Then the system $\phi_n(x)$, $n \in Z^+$, $w(x)$, $x \in (a, b)$, is an orthogonal system with the weight function $w(x)$. If $f(x)$ is a function square integrable on (a, b), then the set of partial sums of its Fourier series with respect to the system converges in the mean to $f(x)$.*

EXERCISES

For further examples see Remark 3.3.2, Exercise 328, formulas (7) and (14) of Section 3.6, and Exercise 641.

237. Prove the Lagrange identity (4).

238. Solve the Sturm–Liouville problems
   a) $y'' + \lambda y = 0$, $y'(0) = y'(\pi) = 0$,
   b) $y'' + \lambda y = 0$, $y(0) = y(1) = 0$,
   c) $y'' + \lambda y = 0$, $y(0) = y'(\pi/2) = 0$.

*239. Solve the Sturm–Liouville problems
   a) $\dfrac{d}{dx}\left(e^{2x}\dfrac{dy}{dx}\right) + e^{2x}(1 + \lambda)y = 0$, $y(0) = y(\pi) = 0$,
   b) $y'' + \lambda y = 0$, $y(0) = y(1) + y'(1) = 0$.

CHAPTER 3

ORTHOGONAL POLYNOMIALS

3.1 THE LEGENDRE POLYNOMIALS

3.1.1 Introductory problem

Find polynomials $P_n(x)$, $n \in N$, such that

1. $P_n(x)$ is of degree n for each $n \in N$,
2. $P_n(x)$, $n \in N$, $x \in (-1, 1)$, is an orthogonal system,
3. $P_n(1) = 1$ for each $n \in N$.

Show that the $P_n(x)$ are uniquely determined by these conditions.

Solution. The sequence of powers x^n, $n \in N$, $x \in (-1, 1)$, is linearly independent (cf. Exercise 232c). Then, by Theorem 2.5.2, there exist polynomials $P_n(x)$ satisfying the conditions 1 and 2. Let $P_n(x)$, $n \in N$, $x \in (-1, 1)$, denote a sequence of such polynomials. Assume $P_0(x) = 1$ for $x \in (-1, 1)$. Fix $n > 0$. Let $F(x)$ denote the polynomial for which $F^{(n)}(x) = P_n(x)$ and

(a) $$F(-1) = F'(-1) = \cdots = F^{(n-1)}(-1) = 0.$$

(Hence $F(x)$ is obtained from $P_n(x)$ by n successive integrations, each time choosing the indefinite integral that has the value zero for $x = -1$.) We also claim that

(b) $$F(1) = F'(1) = \cdots = F^{(n-1)}(1) = 0.$$

To prove (b), suppose that v is an integer, $0 \leqslant v < n$. Then the function x^v is a linear combination of the polynomials $P_0(x)$, $P_1(x)$, ... , $P_v(x)$. The assumption of orthogonality shows that

$$\int_{-1}^{1} x^v P_n(x)\, dx = 0.$$

This gives for $v = 0$:

$$0 = \int_{-1}^{1} x^0 P_n(x)\, dx = \int_{-1}^{1} F^{(n)}(x)\, dx = \left[F^{(n-1)}(x) \right]_{-1}^{1}.$$

It follows that the two last members of (b) are equal. Now suppose that $0 < v < n$, and that the last $v + 1$ members of (b) are equal. Successive integrations by parts give

$$0 = \int_{-1}^{1} x^v P_n(x)\, dx = \int_{-1}^{1} x^v F^{(n)}(x)\, dx$$

45

$$= \left[x^{\nu} F^{(n-1)}(x) \right]_{-1}^{1} - \nu \int_{-1}^{1} x^{\nu-1} F^{(n-1)}(x)\, dx$$

$$= -\nu \int_{-1}^{1} x^{\nu-1} F^{(n-1)}(x)\, dx = \cdots = (-1)^{\nu}(\nu!) \int_{-1}^{1} F^{(n-\nu)}(x)\, dx$$

$$= (-1)^{\nu}(\nu!) \left[F^{(n-\nu-1)}(x) \right]_{-1}^{1}.$$

It follows that the $\nu + 2$ last members of (b) are equal. By induction, all of (b) holds. Using (a) and (b), and observing that $F(x)$ is of degree $2n$, it is seen that $F(x) = c(x^2 - 1)^n$ where c is a constant. The Leibniz formula for the nth derivative of a product then shows that

$$P_n(x) = c\, \frac{d^n}{dx^n} \left[(x+1)^n (x-1)^n \right] = c[(n!)\,(x+1)^n + \cdots],$$

where the three dots denote n terms each having the factor $(x - 1)$. Condition 3 now gives $1 = c(n!)2^n$ and

$$P_n(x) = \frac{1}{2^n(n!)} \frac{d^n}{dx^n} (x^2 - 1)^n, \qquad n \in N. \tag{1}$$

Theorem 2.5.2 shows that the polynomials $P_n(x)$ are uniquely determined.

3.1.2 Definitions

The polynomials obtained in Problem 3.1.1 are called the *Legendre polynomials*. They are denoted by $P_0(x), P_1(x), \ldots$. Formula (1) is called the *Rodrigues formula* for the Legendre polynomials.

3.1.3 Remarks

The result in Problem 3.1.1 shows that $P_n(x)$, $n \in N$, $x \in (-1, 1)$, is an orthogonal system. Table 3.1 shows the computation of a few Legendre polynomials. The six first Legendre polynomials are

$$\begin{aligned}
P_0(x) &= 1, & P_3(x) &= \tfrac{5}{2}x^3 - \tfrac{3}{2}x, \\
P_1(x) &= x, & P_4(x) &= \tfrac{35}{8}x^4 - \tfrac{15}{4}x^2 + \tfrac{3}{8}, \\
P_2(x) &= \tfrac{3}{2}x^2 - \tfrac{1}{2}, & P_5(x) &= \tfrac{63}{8}x^5 - \tfrac{35}{4}x^3 + \tfrac{15}{8}x.
\end{aligned} \tag{2}$$

Table 3.1

n	$(x^2 - 1)^n$	$\dfrac{d^n}{dx^n}(x^2 - 1)^n$	$\dfrac{1}{2^n(n!)}$	$P_n(x)$
0	1	1	1	1
1	$x^2 - 1$	$2x$	$\tfrac{1}{2}$	x
2	$x^4 - 2x^2 + 1$	$12x^2 - 4$	$\tfrac{1}{8}$	$\tfrac{3}{2}x^2 - \tfrac{1}{2}$
3	$x^6 - 3x^4 + 3x^2 - 1$	$120x^3 - 72x$	$\tfrac{1}{48}$	$\tfrac{5}{2}x^3 - \tfrac{3}{2}x$
	$\cdots$	$\cdots$	$\cdots$	$\cdots$

It is seen that $P_n(x)$ contains only even powers of x if n is even, only odd powers if n is odd. Further

$$P_n(1) = 1, \quad P_n(-1) = (-1)^n, \quad n \in N. \tag{3}$$

Figure 3.1 shows the first five Legendre polynomials.

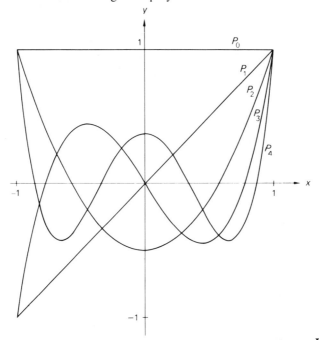

Figure 3.1

Let n be a given positive integer. It is seen that the function $F(x) = (x^2 - 1)^n$ and its $n - 1$ first derivatives all have zeros at $x = 1$ and at $x = -1$. Successive applications of Rolle's theorem shows that $F'(x)$ has at least one zero in the open interval $(-1, 1)$, that $F''(x)$ has at least two zeros in the same interval, etc. It follows that the Legendre polynomial $P_n(x)$ has exactly n zeros in the open interval $(-1, 1)$.

EXERCISES

301. Verify the expression for $P_4(x)$ in (2).
302. Find the maximum and minimum points of the curves $y = P_2(x)$ and $y = P_3(x)$.
303. Show by means of the Rodrigues formula that

$$P_n'(1) = (-1)^{n-1} P_n'(-1) = \frac{n(n+1)}{2}.$$

*304. a) Compute $P_n(0)$.
    b) Does the limit $\lim_{n \to \infty} P_n(0)$ exist?
    Hint for (b). Use the Stirling formula: $n! \approx (2\pi n)^{1/2} n^n e^{-n}$, where $f(n) \approx g(n)$ means
    that $f(n)/g(n) \to 1$ as $n \to \infty$.

305. Solve Problem 3.1.1 after having replaced the interval $(-1, 1)$ by (a, b) and $P_n(1)$
    by $P_n(b)$.

3.2 LEGENDRE SERIES

This section (3.2) is devoted to a study of Fourier series with respect to the orthogonal
system of Problem 3.1.1. We first give a recursion formula, formula (1) below, by means
of which the polynomial $P_n(x)$ can be computed for a given $n \geqslant 2$, if the two immediately
preceding polynomials are known. This formula will enable us to compute the norms of
the Legendre polynomials.

3.2.1 Lemma (giving a recursion formula for the Legendre polynomials).
Suppose that $P_{n-2}(x)$, $P_{n-1}(x)$, $P_n(x)$ are Legendre polynomials. Then for all x,

$$nP_n(x) - (2n - 1)xP_{n-1}(x) + (n - 1)P_{n-2}(x) = 0. \tag{1}$$

Proof. The coefficient of x^n in the polynomial

(a) $$nP_n(x) - (2n - 1)xP_{n-1}(x)$$

is

$$n\frac{1}{2^n}\frac{1}{n!}\frac{(2n)!}{n!} - (2n - 1)\frac{1}{2^{n-1}}\frac{1}{(n-1)!}\frac{(2n-2)!}{(n-1)!}$$
$$= \left[n\frac{1}{2n}\frac{2n(2n-1)}{n} - (2n - 1)\right]\frac{1}{2^{n-1}}\frac{1}{(n-1)!}\frac{(2n-2)!}{(n-1)!} = 0.$$

Then (a) is a polynomial of degree $n - 2$, and there exist constants $c_0, c_1, \ldots,$
c_{n-2} such that for all x

(b) $$nP_n(x) - (2n - 1)xP_{n-1}(x) = c_0 P_0(x) + \cdots + c_{n-2} P_{n-2}(x).$$

Let v denote one among the numbers $0, 1, \ldots, n - 3$. Multiply each member of
(b) by $P_v(x)$, and integrate over the interval $(-1, 1)$. Then, by orthogonality,

$$-(2n - 1)\int_{-1}^{1} xP_v(x)\,P_{n-1}(x)\,dx = c_v \int_{-1}^{1} P_v^2(x)\,dx.$$

Here $xP_v(x)$ is polynomial of degree $v + 1$. Thus the left member is zero by
orthogonality. Hence $c_v = 0$ for $v = 0, 1, \ldots, n - 3$. The equation (b) then
gives for all x

$$nP_n(x) - (2n - 1)xP_{n-1}(x) = c_{n-2}P_{n-2}(x).$$

Here set $x = 1$:

$$n - (2n - 1) = c_{n-2}, \qquad c_{n-2} = -(n - 1).$$

Formula (1) follows.

3.2.2 Lemma. *The square of the norm of a Legendre polynomial is*

$$\|P_n\|^2 = \frac{2}{2n+1}, \qquad n \in N. \tag{2}$$

Proof. Formula (2) is immediately verified for $n = 0$ and $n = 1$. Suppose $n \geqslant 2$. Lemma 3.2.1 gives for all x

$$nP_n(x) - (2n-1)xP_{n-1}(x) + (n-1)P_{n-2}(x) = 0,$$
$$(n+1)P_{n+1}(x) - (2n+1)xP_n(x) + nP_{n-1}(x) = 0.$$

Now multiply both members by $P_n(x)$ and $P_{n-1}(x)$ respectively and integrate over the interval $(-1, 1)$:

$$n \int_{-1}^{1} P_n^2(x)\, dx = (2n-1) \int_{-1}^{1} xP_{n-1}(x)P_n(x)\, dx.$$

$$n \int_{-1}^{1} P_{n-1}^2(x)\, dx = (2n+1) \int_{-1}^{1} xP_{n-1}(x)P_n(x)\, dx.$$

These equations show that

$$(2n+1) \int_{-1}^{1} P_n^2(x)\, dx = (2n-1) \int_{-1}^{1} P_{n-1}^2(x)\, dx = \cdots = 3 \int_{-1}^{1} P_1^2(x)\, dx = 2.$$

Now the assertion follows.

3.2.3 Legendre series

By Problem 3.1.1 the polynomials $P_n(x)$, $n \in N$, and the interval $(-1, 1)$ form an orthogonal system. Let $f(x)$ be a complex-valued function with domain $(-1, 1)$, integrable on its domain. By Definition 2.2.3, $f(x)$ has a Fourier series with respect to this system:

$$f(x) \sim \sum_{n=0}^{\infty} c_n P_n(x), \tag{3}$$

where, using Lemma 3.2.2,

$$c_n = \frac{2n+1}{2} \int_{-1}^{1} f(x)P_n(x)\, dx, \qquad n \in N. \tag{4}$$

The numbers (4) are called the *Legendre coefficients* of $f(x)$, and the series in (3) is called the *Legendre series* of $f(x)$. Corollary 3.2.5 below shows that this series converges in the mean to $f(x)$, if $f(x)$ is square integrable on $(-1, 1)$. Theorems on pointwise convergence can also be proved (see [9], Theorem 11.8, or [14], Section 3.14).

3.2.4 Theorem. *The orthogonal system* $P_n(x)$, $n \in N$, $x \in (-1, 1)$, *where* $P_n(x)$ *denotes the Legendre polynomials, has the Parseval property.*

Proof. Suppose that $\varepsilon > 0$ and that $f(x)$ is a complex-valued function, square integrable on the interval $(-1, 1)$. By Exercise 127 there is a function $g(x)$, continuous on the interval $[-1, 1]$, such that

$$\| f(x) - g(x) \|^2 < \frac{\varepsilon}{4}.$$

By Exercise 129 there is a nonnegative integer n and a polynomial $p_n(x)$ of degree n such that

$$\| g(x) - p_n(x) \|^2 < \frac{\varepsilon}{4}.$$

The parallelogram law (see Example 2.3.7) now shows that

$$\| f(x) - p_n(x) \|^2 < \varepsilon.$$

Let $s_n(x)$ be the nth partial sum of the Legendre series of $f(x)$. Then, by Theorem 2.3.1,

$$\| f(x) - s_n(x) \|^2 < \varepsilon.$$

Here, by the same theorem, the index n can be replaced by any larger index. Then $\| f(x) - s_n(x) \| \to 0$ as $n \to \infty$, and Corollary 2.3.2 gives the assertion.

We rephrase part of the last sentence as a corollary.

3.2.5 Corollary. *If* $f(x)$ *is a complex-valued function, square integrable on the interval* $(-1, 1)$, *then the sequence of partial sums* $s_n(x)$ *of the Legendre series of* $f(x)$ *converges in the mean to* $f(x)$.

3.2.6 Example
Consider the odd function $f(x)$ with domain $(-1, 1)$, for which $f(x) = 1$ for $0 < x < 1$.
   a) Find the partial sum $s_3(x) = \sum_{\nu=0}^{3} c_\nu P_\nu(x)$ of the Legendre series of $f(x)$.
   b) Find a polynomial $p(x) = \sum_{\nu=0}^{3} a_\nu x^\nu$ of third or lower degree such that the norm $\| f(x) - p(x) \|$ is as small as possible.

Solution. a) Formula (4) shows that $c_0 = c_2 = 0$, since $f(x)$ is an odd function. The same formula gives

$$c_1 = 2 \cdot \tfrac{3}{2} \int_0^1 x \, dx = \tfrac{3}{2}, \qquad c_3 = 2 \cdot \tfrac{7}{2} \int_0^1 (\tfrac{5}{2}x^3 - \tfrac{3}{2}x) \, dx = -\tfrac{7}{8}.$$

Hence

$$s_3(x) = \tfrac{3}{2}P_1(x) - \tfrac{7}{8}P_3(x).$$

b) Theorem 2.3.1 shows that

$$p(x) = s_3(x) = \tfrac{3}{2}x - \tfrac{7}{8}(\tfrac{5}{2}x^3 - \tfrac{3}{2}x) = \tfrac{45}{16}x - \tfrac{35}{16}x^3.$$

EXERCISES

306. Verify the expressions for $P_4(x)$ and $P_5(x)$ in formula (2) of Section 3.1.3, using the recursion formula (1).

307. Expand the function $f(x) = x^2$, $-1 < x < 1$, in a Legendre series.

308. Solve the problem in 3.2.6 after having replaced the function $f(x)$ by the function $g(x) = |x|$, $-1 < x < 1$.

309. Find an orthonormal system corresponding to the system of Problem 3.1.1; normalize by choosing positive values at the point $x = 1$.

310. Show that $\int_{-1}^{1} |P_n(x)| \, dx < (2/n)^{1/2}$ for each $n \geqslant 1$.
   Hint. Use the Schwarz inequality.

*311. Find three polynomials $p_0(x), p_1(x), p_2(x)$ of degrees 0, 1, 2 respectively, which together with the interval $(-1, 1)$ form an orthonormal system; normalize by choosing positive values at the point $x = 1$. Show that each polynomial $p(x)$ of second degree can be represented as a linear combination of these three polynomials: $p(x) = \sum_{v=0}^{2} c_v p_v(x)$, and that the coefficients c_v are uniquely determined by $p(x)$.

312. Use the Rodrigues formula and the recursion formula (1) to show that for all x:
   a) $P'_{n+1}(x) - xP'_n(x) = (n+1)P_n(x)$,
   b) $P'_{n+1}(x) - P'_{n-1}(x) = (2n+1)P_n(x)$,
   c) $(x^2 - 1)P'_n(x) = nxP_n(x) - nP_{n-1}(x)$.

313. Suppose that $P_{n-1}(x)$, $P_n(x)$, $P_{n+1}(x)$ are three successive Legendre polynomials and that x_0 is a point in the open interval $(-1, 1)$, such that $P_{n-1}(x_0) = P_{n+1}(x_0)$. Show that $P_n(x)$ has a maximum or a minimum for $x = x_0$.

314. Is the system $P_{2n}(x)$, $n \in N$, $x \in (0, 1)$, a complete orthogonal system?

3.3 THE LEGENDRE DIFFERENTIAL EQUATION.
THE GENERATING FUNCTION OF THE LEGENDRE POLYNOMIALS

We shall show that each Legendre polynomial satisfies a certain differential equation, called the Legendre differential equation.

We shall also show that a certain series expansion produces the Legendre polynomials. The function that is expanded is called the generating function of the Legendre polynomials. In our discussion of this function it is assumed that the reader has some knowledge of functions analytic in a domain (see [1], especially the sections beginning on pp. 69 and 177).

3.3.1 Theorem (giving the Legendre differential equation). *The Legendre polynomial $P_n(x)$ is a solution of the differential equation*

$$(1 - x^2)y'' - 2xy' + n(n+1)y = 0, \qquad n \in N. \tag{1}$$

Proof. Set $F(x) = (x^2 - 1)^n$. Then $F'(x) = 2nx(x^2 - 1)^{n-1}$ and $2nxF(x) = (x^2 - 1)F'(x)$.

Differentiation $n + 1$ times, using the Leibniz formula, gives

$$2nxF^{(n+1)}(x) + 2n(n + 1)F^{(n)}(x)$$
$$= (x^2 - 1)F^{(n+2)}(x) + 2(n + 1)xF^{(n+1)}(x) + 2\frac{(n + 1)n}{2}F^{(n)}(x),$$
$$(1 - x^2)F^{(n+2)}(x) - 2xF^{(n+1)}(x) + n(n + 1)F^{(n)}(x) = 0.$$

Here divide both members by $2^n(n!)$, and apply the Rodrigues formula. Then (1) is obtained.

For the general solution of the Legendre differential equation, see [18], p. 49.

3.3.2   Remark

The differential equation (1) reminds one of the differential equation in a Sturm–Liouville problem (see Definition 2.6.2). There are, however, two differences. The coefficient of y'' in (1) is not positive at the end points of the interval $[-1, 1]$. Boundary conditions such as (2b, c) in Definition 2.6.2 cannot be associated with (1) in a reasonable way. However, the definitions of eigenvalue and eigenfunction in Definition 2.6.2 are applicable to the boundary-value problem

$$\frac{d}{dx}\left[(1 - x^2)\frac{dy}{dx}\right] + \lambda y = 0, \qquad x \in [-1, 1], \tag{2a}$$

$$\lim_{x \to 1} y(x) \quad \text{and} \quad \lim_{x \to -1} y(x) \text{ exist.} \tag{2b}$$

It is seen that the numbers $\lambda_n = n(n + 1)$, $n \in N$, are eigenvalues for the problem (2) and that the Legendre polynomials are corresponding eigenfunctions. (They are in fact *the* eigenvalues and, up to nonzero factors, *the* eigenfunctions; see [18], p. 109.)

3.3.3   Convention

Suppose that x is a real number such that $-1 \leqslant x \leqslant 1$, and that t is a complex variable. The polynomial $1 - 2xt + t^2$ has its zeros on the circle $|t| = 1$. Consider the expression

(a)
$$\frac{1}{\sqrt{1 - 2xt + t^2}}.$$

For each t such that $|t| < 1$ it has two values (with the sum zero). There is, however, a branch of the expression (a) that is analytic for $|t| < 1$ and that has the value 1 for $t = 0$. In the following theorem we let (a) denote this branch.

3.3.4   Theorem (giving the generating function of the Legendre polynomials). *Suppose that x is a real number in the interval $[-1, 1]$ and that t is a complex* *variable in the disk* $|t| < 1$. *Then, adhering to the above convention,*

$$\frac{1}{\sqrt{1 - 2xt + t^2}} = \sum_{n=0}^{\infty} P_n(x)t^n. \tag{3}$$

Proof. The left member of (3) is equal to the sum of its Taylor series:

$$\frac{1}{\sqrt{1 - 2xt + t^2}} = \sum_{n=0}^{\infty} Q_n(x)t^n,$$

where the notation $Q_n(x)$ indicates that the coefficients of the power series depend on x. Computation of the values of the left member of (3) and its derivative for $t = 0$ shows that $Q_0(x) = 1$ and $Q_1(x) = x$. Hence

(b) $\qquad\qquad Q_n(x) = P_n(x) \qquad$ for $n = 0$ and $n = 1$,

where $P_n(x)$ denotes Legendre polynomials. The following equations also hold (differentiation term by term is legitimate):

$$\frac{x - t}{(1 - 2xt + t^2)^{3/2}} = Q_1(x) + 2Q_2(x)t + \cdots + nQ_n(x)t^{n-1} + \cdots,$$

$$(x - t)(Q_0(x) + Q_1(x)t + \cdots + Q_n(x)t^n + \cdots)$$
$$= (1 - 2xt + t^2)(Q_1(x) + 2Q_2(x)t + \cdots + nQ_n(x)t^{n-1} + \cdots).$$

Suppose $n > 1$. The coefficients of t^{n-1} in the two members are equal:

$$xQ_{n-1}(x) - Q_{n-2}(x) = nQ_n(x) - 2x(n - 1)Q_{n-1}(x) + (n - 2)Q_{n-2}(x),$$

and hence

(c) $\qquad nQ_n(x) - (2n - 1)xQ_{n-1}(x) + (n - 1)Q_{n-2}(x) = 0 \qquad$ for $n > 1$.

The equations (b) and (c) and the recursion formula for the Legendre polynomials show that $Q_n(x) = P_n(x)$ for each $n \in N$. This gives the assertion.

EXERCISES

315. Find a polynomial $p(x)$ such that $p(0) = 3$ and $(1 - x^2)p''(x) - 2xp'(x) + 20p(x) = 0$.

316. a) Suppose that $P_m(x)$ and $P_n(x)$ are Legendre polynomials such that $m \neq n$. Show that $P_m(x)$ is orthogonal to $(d/dx)[(x^2 - 1)P_n'(x)]$ on the interval $(-1, 1)$.
  b) Suppose that $P_m(x)$ and $P_n(x)$ are Legendre polynomials such that $n > 0, m \neq n + 1$, and $m \neq n - 1$. Show that $P_m(x)$ is orthogonal to $(x^2 - 1)P_n'(x)$ on the interval $(-1, 1)$. *Hint* for (b). Apply the result in Exercise 312c.

317. Check part of Theorem 3.3.4 by computing the values of the left member of (3) and its two first derivatives for $t = 0$.

*318. Sum the series $\sum_{n=0}^{\infty} (n + 1)^{-1} x^{n+1} P_n(x)$ for $|x| < 1$.

3.4 THE TCHEBYCHEFF POLYNOMIALS

We shall study a set of polynomials named after the Russian mathematician Tchebycheff (the transliteration Chebichev and other spellings of his name are often used). He found them as the solutions of the following problem (see [7], pp. 111–113; cf. also Exercises 319 and 320): which real-valued polynomial $T_n(x)$ of given degree n, such that $|T_n(x)| \leqslant 1$ for $|x| \leqslant 1$, has the largest coefficient for x^n? We shall give here another approach to the Tchebycheff polynomials.

In this section and the next one (3.4 and 3.5), $w(x)$ will always denote the weight function $w(x) = (1 - x^2)^{-1/2}, |x| < 1$ (see Fig. 3.2a).

3.4.1 Introductory problem

Find polynomials $T_n(x)$, $n \in N$, such that

1. $T_n(x)$ is of degree n for each $n \in N$,
2. $T_n(x)$, $n \in N$, $w(x)$, $x \in (-1, 1)$, is an orthogonal system with the weight function $w(x) = (1 - x^2)^{-1/2}$ (cf. Definition 2.6.4),
3. $T_n(1) = 1$ for each $n \in N$.

Show that the $T_n(x)$ are uniquely determined by these conditions.

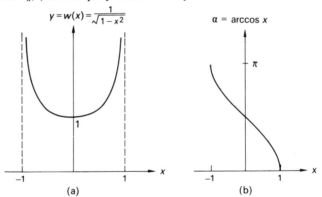

Figure 3.2

Solution. It is readily verified that the sequence of functions $x^n\sqrt{w(x)}$, $n \in N$, $x \in (-1, 1)$, is linearly independent. Then, by Theorem 2.5.2, there exists at most one sequence of polynomials satisfying the conditions 1, 2, 3. Suppose that such a sequence exists, and denote it by $T_n(x)$, $n \in N$. Let m and n be two nonnegative integers. Then the substitution (see Figure 3.2b)

(a) $\alpha = \arccos x$

shows that

$$0 = \int_{-1}^{1} \frac{T_m(x) T_n(x)}{\sqrt{1 - x^2}} \, dx = \int_{0}^{\pi} T_m(\cos \alpha) T_n(\cos \alpha) \, d\alpha.$$

Now observe the following three facts:

1′. Comparison of the real parts of the two members of de Moivre's formula (expand the right member by the binomial theorem),

(b) $\cos n\alpha + i \sin n\alpha = (\cos \alpha + i \sin \alpha)^n$,

shows that $\cos n\alpha$ is a polynomial of degree n in terms of $\cos \alpha$.

2′. Example 2.2.5 shows that

$$m, n \in N \text{ and } m \neq n \quad \Rightarrow \quad \int_0^\pi \cos m\alpha \cos n\alpha \, d\alpha = 0.$$

3′. For $\alpha = 0$ and each $n \in N$, $\cos n\alpha = 1$.

These facts show that $T_n(\cos \alpha) = \cos n\alpha$ for each n. It follows that

$$T_n(x) = \cos (n \arccos x), \qquad n \in N, \quad x \in [-1, 1], \tag{1}$$

is a solution of Problem 3.4.1, and that this solution is unique.

3.4.2 Definition

The polynomials obtained in Problem 3.4.1 are called the *Tchebycheff polynomials*. They are denoted by $T_0(x)$, $T_1(x), \ldots$. Formula (1) gives the Tchebycheff polynomials.

3.4.3 Remarks

The result in Problem 3.4.1 shows that $T_n(x)$, $n \in N$, $w(x)$, $x \in (-1, 1)$, is an orthogonal system with the weight function $w(x)$ of Figure 3.2a. Formula (b) gives

$$\cos n\alpha = \cos^n\alpha - \binom{n}{2}\cos^{n-2}\alpha \sin^2\alpha + \binom{n}{4} \cos^{n-4}\alpha \sin^4\alpha - \cdots,$$

whence by (a) and (1),

$$T_n(x) = x^n - \binom{n}{2} x^{n-2}(1 - x^2) + \binom{n}{4} x^{n-4}(1 - x^2)^2 - \cdots, \tag{2}$$

the three dots indicating a finite number of terms (cf. Exercise 321). The six first Tchebycheff polynomials are

$$\begin{aligned}
T_0(x) &= 1, & T_3(x) &= 4x^3 - 3x, \\
T_1(x) &= x, & T_4(x) &= 8x^4 - 8x^2 + 1, \\
T_2(x) &= 2x^2 - 1, & T_5(x) &= 16x^5 - 20x^3 + 5x.
\end{aligned} \tag{3}$$

It is seen that $T_n(x)$ contains only even powers of x if n is even, only odd powers if n is odd. Further

$$T_n(1) = 1, \, T_n(-1) = (-1)^n, \qquad n \in N. \tag{4}$$

Figure 3.3 shows the first five Tchebycheff polynomials.

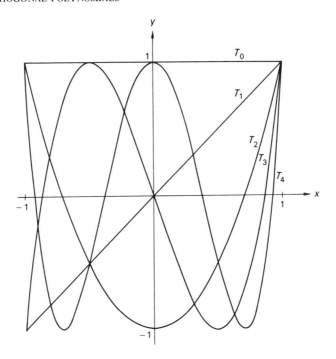

Figure 3.3

Formula (1) shows that the zeros of $T_n(x)$ are those values of x for which $(n \arccos x)$ is an odd multiple of $\pi/2$. Hence $T_n(x)$ has exactly n zeros:

$$x = \cos \frac{2v - 1}{2n} \pi, \qquad v \in \{1, 2, \ldots, n\}. \tag{5}$$

EXERCISES

319. Show that $|T_n(x)| \leqslant 1$ for $|x| \leqslant 1$.

320. Find the coefficient of x^n in $T_n(x)$.

321. Find the last term in the right member of (2).

322. Evaluate the integral $\int_{-1}^{1} T_n(x)\, dx$.
     Hint. Use the substitution (a).

323. Deduce the following recursion formula for the Tchebycheff polynomials:

$$T_{n+1}(x) - 2xT_n(x) + T_{n-1}(x) = 0.$$

Hint. The result follows from:

$$2x\,T_n(x) = 2\cos{(\arccos x)}\cos{(n\arccos x)},$$
$$2\cos\alpha\cos\beta = \cos{(\alpha+\beta)} + \cos{(\alpha-\beta)}.$$

324. Verify the expressions for $T_0(x), \ldots, T_5(x)$ in (3).
   Hint. Use (1) for the two first expressions and the recursion formula in Exercise 323 for the remaining ones.

325. For which values of x in the open interval $(-1, 1)$ has $T_n(x)$ an extremum?

326. Show that $T_n(x)$ satisfies the differential equation

$$(1 - x^2)y'' - xy' + n^2y = 0,$$

   called the *Tchebycheff differential equation.*

*327. Show that if x is a real number in the interval $[-1, 1]$ and if t is a complex variable in the disk $|t| < 1$, then

$$\frac{1 - xt}{1 - 2xt + t^2} = \sum_{n=0}^{\infty} T_n(x)t^n.$$

   Here the left member is called the *generating function* of the Tchebycheff polynomials.

328. Formulate a remark on the Tchebycheff polynomials, analogous to Remark 3.3.2. In particular:
   a) Give a differential equation similar to (2a) in Definition 2.6.2.
   b) Give boundary conditions for $x = 1$ and $x = -1$.
   c) Give a sequence of eigenvalues.
   d) Give corresponding eigenfunctions.

3.5 TCHEBYCHEFF SERIES

3.5.1 Definitions

Consider the system

$$T_n(x),\ n \in N, \qquad w(x),\ x \in (-1, 1), \tag{1}$$

consisting of the Tchebycheff polynomials, the weight function $w(x) = (1 - x^2)^{-1/2}$, and the interval $(-1, 1)$. By the result in Problem 3.4.1, the system (1) is an orthogonal system with the weight function $w(x)$. We apply Definitions 2.6.4 to this system. The square of the norm of a Tchebycheff polynomial is readily computed by use of the substitution $\alpha = \arccos x$:

$$\|T_n\|^2 = \int_{-1}^{1} T_n^2(x)w(x)\,dx = \int_{-1}^{1} \frac{\cos^2(n\arccos x)}{\sqrt{1 - x^2}}\,dx = \int_{0}^{\pi} \cos^2 n\alpha\,d\alpha,$$

and hence

$$\|T_0\|^2 = \pi \quad \text{and} \quad \|T_n\|^2 = \frac{\pi}{2} \quad \text{for } n > 0. \tag{2}$$

Suppose that $f(x)$ is a complex-valued function and that the product $f(x)w(x)$ is integrable on $(-1, 1)$. Then, by Definition 2.6.4, $f(x)$ has Fourier coefficients c_n with respect to the system (1):

$$c_0 = \frac{1}{\pi} \int_{-1}^{1} f(x)w(x)\, dx, \tag{3a}$$

$$c_n = \frac{2}{\pi} \int_{-1}^{1} f(x)T_n(x)w(x)\, dx \quad \text{for } n > 0, \tag{3b}$$

and a Fourier series with respect to the system (1):

$$f(x) \sim \sum_{n=0}^{\infty} c_n T_n(x). \tag{4}$$

The numbers (3) are called the *Tchebycheff coefficients* of $f(x)$, and the series in (4) is called the *Tchebycheff series* of $f(x)$.

In forming the formulas (3) we assumed the product $f(x)w(x)$ to be integrable on $(-1, 1)$. This assumption is necessary: It is verified that the function $f(x) = (1 - x^2)^{-1/2}$, $-1 < x < 1$, is integrable on $(-1, 1)$, but that it has no Tchebycheff coefficients.

We shall state two theorems. In the first theorem it is shown that the series (4) is closely connected with a certain cosine series. The second theorem deals with convergence in the mean. Theorems on pointwise convergence can also be proved (see Exercise 332). The two theorems can be proved by use of the substitution $\alpha = \arccos x$; the proofs are referred to Exercise 331.

3.5.2 Theorem. *Suppose that the complex-valued function $f(x)$ has the Tchebycheff coefficients c_n, $n \in N$. Set $g(\alpha) = f(\cos \alpha)$ for $0 < \alpha < \pi$. Denote by a_n, $n \in N$, the cosine coefficients of the function $g(\alpha)$ (i.e. $g(\alpha) \sim \frac{1}{2}a_0 + \sum_{n=1}^{\infty} a_n \cos n\alpha$). Then*

$$c_0 = \tfrac{1}{2}a_0 \quad \text{and} \quad c_n = a_n \quad \text{for} \quad n > 0. \tag{5}$$

3.5.3 Theorem. *Suppose that the functions $f(x)$ and $g(\alpha)$ satisfy the hypotheses of Theorem 3.5.2 and that $g(\alpha)$ is square integrable on the interval $(0, \pi)$. Then the sequence of partial sums $s_n(x)$ of the Tchebycheff series of $f(x)$ converges in the mean to $f(x)$, i.e.*

$$\| f(x) - s_n(x) \|^2 = \int_{-1}^{1} \Big| f(x) - \sum_{v=0}^{n} c_v T_v(x) \Big|^2 w(x)\, dx \to 0 \quad \text{as} \quad n \to \infty. \tag{6}$$

3.5.4 Example

Consider the function $f(x)$ defined by:

$$f(x) = 1 \quad \text{for} \quad 0 < x < 1, \qquad f(x) = 0 \quad \text{for} \quad -1 < x < 0.$$

Expand $f(x)$ in a Tchebycheff series.

Solution. Set $g(\alpha) = f(\cos \alpha)$ for $0 < \alpha < \pi$. Then $g(\alpha) = 1$ for $0 < \alpha < \pi/2$, $g(\alpha) = 0$ for $\pi/2 < \alpha < \pi$. The coefficients of the cosine series of $g(\alpha)$ are

$$a_0 = \frac{2}{\pi} \int_0^{\pi/2} 1 \, d\alpha = 1,$$

$$a_n = \frac{2}{\pi} \int_0^{\pi/2} \cos n\alpha \, d\alpha = \frac{2}{n\pi} \sin \frac{n\pi}{2} \quad \text{for} \quad n > 0.$$

Then, by Theorem 3.5.2,

$$f(x) \sim \frac{1}{2} + \frac{2}{\pi} \sum_{n=0}^{\infty} \frac{(-1)^n}{2n+1} T_{2n+1}(x).$$

EXERCISES

329. Consider the function $f(x) = 1 + x + x^2$, $|x| \leqslant 1$.
     a) Express $f(x)$ as a linear combination of the first three Tchebycheff polynomials.
     b) Expand $f(x)$ in a Tchebycheff series.
330. Expand the function $f(x) = \arccos x$ in a Tchebycheff series.
*331. Prove Theorems 3.5.2 and 3.5.3.

332. State and prove a theorem for the Tchebycheff series analogous to Theorem 1.4.4.

3.6 THE HERMITE POLYNOMIALS. THE LAGUERRE POLYNOMIALS

So far in this chapter we have studied polynomials having orthogonality properties on a finite interval (a, b) of the x-axis. (For the most part the specific interval $(-1, 1)$ has been considered. Exercise 305 shows that this restriction is not essential.)

     We now give a short exposition on two sets of polynomials having orthogonality properties on infinite intervals. Proofs are omitted. Definitions of fundamental concepts (integrable, orthogonal, complete) in the new situation are also omitted. Definitions and proofs can be studied in [14], Chapter 4, or in [18], Sections 24, 51 and 54.

3.6.1 The Hermite polynomials

The sequence of functions $x^n e^{-x^2/2}$, $n \in N$, $x \in (-\infty, \infty)$, is linearly independent. The Gram-Schmidt process produces a system $H_n(x)$, $n \in N$, e^{-x^2}, $x \in (-\infty, \infty)$, that is an orthogonal system with the weight function e^{-x^2}. Here each $H_n(x)$ is a polynomial of degree n. The $H_n(x)$ are called Hermite polynomials. The following identities hold:

$$H_n(x) = (-1)^n e^{x^2} \frac{d^n}{dx^n} e^{-x^2}, \qquad n \in N. \tag{1}$$

The first two polynomials are

$$H_0(x) = 1, \qquad H_1(x) = 2x. \tag{2}$$

The polynomials $H_2(x)$, $H_3(x)$, ... can be computed by the recursion formula

$$H_{n+1}(x) - 2x H_n(x) + 2n H_{n-1}(x) = 0. \tag{3}$$

The polynomial $H_n(x)$ satisfies the Hermite differential equation:

$$y'' - 2xy' + 2ny = 0. \tag{4}$$

The polynomials have a generating function:

$$e^{-t^2 + 2tx} = \sum_{n=0}^{\infty} \frac{H_n(x)}{n!} t^n. \tag{5}$$

The square of the norm of $H_n(x)$ is

$$\|H_n\|^2 = 2^n(n!) \sqrt{\pi}. \tag{6}$$

For the problem:

$$\frac{d}{dx}\left(e^{-x^2} \frac{dy}{dx}\right) + \lambda e^{-x^2} y = 0, \qquad x \in (-\infty, \infty), \tag{7a}$$

$$\text{there exists } m \in N \text{ such that } y(x)x^{-m} \to 0 \text{ as } |x| \to \infty, \tag{7b}$$

the eigenvalues are $\lambda_n = 2n$, $n \in N$, and the corresponding eigenfunctions are $H_n(x)$, $n \in N$, up to nonzero factors.

3.6.2 The Laguerre polynomials.

The sequence of functions $x^n e^{-x/2}$, $n \in N$, $x \in (0, \infty)$, is linearly independent. The Gram-Schmidt process produces a system $L_n(x)$, $n \in N$, e^{-x}, $x \in (0, \infty)$, that is orthogonal with the weight function e^{-x}. Here each $L_n(x)$ is a polynomial of degree n. The $L_n(x)$ are called Laguerre polynomials. The following identities hold:

$$L_n(x) = \frac{1}{n!} e^x \frac{d^n}{dx^n} (x^n e^{-x}), \qquad n \in N. \tag{8}$$

The two first polynomials are

$$L_0(x) = 1, \qquad L_1(x) = -x + 1. \tag{9}$$

The polynomials $L_2(x)$, $L_3(x)$, ... can be computed by the recursion formula:

$$(n + 1)L_{n+1}(x) - (2n + 1 - x)L_n(x) + nL_{n-1}(x) = 0. \tag{10}$$

The polynomial $L_n(x)$ satisfies the Laguerre differential equation:

$$xy'' + (1 - x)y' + ny = 0. \tag{11}$$

The polynomials have a generating function:

$$\frac{1}{1 - t} e^{-xt/(1-t)} = \sum_{n=0}^{\infty} L_n(x)t^n, \qquad |t| < 1. \tag{12}$$

The norm of $L_n(x)$ is

$$\|L_n\| = 1. \tag{13}$$

For the problem:

$$\frac{d}{dx}\left(xe^{-x}\frac{dy}{dx}\right) + \lambda e^{-x}y = 0, \qquad \lim_{x\to 0} y(x) \text{ exists,} \qquad \text{(14a, b)}$$

$$\text{there exists } m \in N \text{ such that } y(x)x^{-m} \to 0 \text{ as } x \to \infty, \qquad \text{(14c)}$$

the eigenvalues are $\lambda_n = n$, $n \in N$, and the corresponding eigenfunctions are $L_n(x)$, $n \in N$, up to nonzero factors.

EXERCISE

*333. Show that the sequence of functions introduced in the first sentence of Section 3.6.1 is linearly independent, and do the same for Section 3.6.2.

CHAPTER 4

FOURIER TRANSFORMS

4.1 INFINITE INTERVAL OF INTEGRATION

We want to extend the concepts of integrable function and integral so that the interval of integration can be infinite. We shall use the following definitions.

4.1.1 Definitions

Consider a complex-valued function $f(x)$, integrable on each finite subinterval of the infinite interval (a, ∞). First suppose that its integral on the interval (a, b) tends to a limit as $b \to \infty$. We define the *improper integral* of $f(x)$ on (a, ∞), denoted by $\int_a^\infty f(x)\,dx$, by

$$\int_a^\infty f(x)\,dx = \lim_{b \to \infty} \int_a^b f(x)\,dx. \tag{1}$$

(For the next definition, observe that $|f(x)|$ is integrable on each finite subinterval of (a, ∞); cf. Exercise 203.) Secondly, suppose that $|f(x)|$ has an improper integral on (a, ∞). It follows that the limit in (1) exists (Exercise 401). In this case we say that $f(x)$ is *integrable* on (a, ∞), and we define its *integral* on (a, ∞), also denoted by $\int_a^\infty f(x)\,dx$, by (1). It should be observed that a function can have an improper integral on (a, ∞) without being integrable on this infinite interval (cf. Exercise 402). The corresponding three notions for the interval $(-\infty, a)$ are introduced analogously.

Now suppose that the function $f(x)$ is integrable on each finite interval. If its integral on the interval $(-a, a)$ tends to a limit as $a \to \infty$, we define the *improper integral* of $f(x)$ on $(-\infty, \infty)$, denoted $\int_{-\infty}^\infty f(x)\,dx$, by

$$\int_{-\infty}^\infty f(x)\,dx = \lim_{a \to \infty} \int_{-a}^a f(x)\,dx. \tag{2}$$

If $|f(x)|$ has an improper integral on $(-\infty, \infty)$, we say that $f(x)$ is *integrable* on $(-\infty, \infty)$ and we define its *integral* on $(-\infty, \infty)$, also denoted by $\int_{-\infty}^\infty f(x)\,dx$, by (2).

The definitions (1) and (2) have been introduced here for practical reasons. A reader who dislikes to use the notations in the left members of (1) and (2) when the integrand is not integrable on the interval of integration, can use the notations in the right members at certain places in the following text. Also observe that what was said in the comments to formula (8c) of Definition 2.2.3, can here be repeated with some obvious changes in the formulations.

A word of warning: usual rules for handling integrals are not all valid for improper integrals. An instance is given in Exercise 405, where a substitution changes the value of an improper integral.

EXERCISES

*401. Show that if $f(x)$ is integrable on each finite subinterval of (a, ∞) and $\lim_{b \to \infty} \int_a^b |f(x)| \, dx$ exists, then $\lim_{b \to \infty} \int_a^b f(x) \, dx$ exists.

  Hint. Set $F(b) = \int_a^b f(x) \, dx$. Suppose that $\varepsilon > 0$, and that m and n are positive integers such that $a < m < n$. Then, for m sufficiently large, by the result in Exercise 203

$$|F(n) - F(m)| = \left| \int_m^n f(x) \, dx \right| \leqslant \int_m^\infty |f(x)| \, dx < \varepsilon.$$

  It follows (see [13], Theorem 3.11) that $\lim_{n \to \infty} F(n)$ exists. Deduce that $\lim_{b \to \infty} F(b)$ exists.

402. Consider the function $f(x) = x^{-\alpha} \sin x$, $0 < x < \infty$, where α is given a positive number. Find the positive values of α for which (a) $f(x)$ is integrable on $(0, \infty)$, (b) $f(x)$ has an improper integral on $(0, \infty)$.

403. Consider the function $f(x) = (1 + x)/(1 + x^2)$. (a) Is $f(x)$ integrable on $(-\infty, \infty)$? (b) Has $f(x)$ an improper integral on $(-\infty, \infty)$? (c) Does the integral $\int_a^b f(x) dx$ tend to a limit as $a \to -\infty$ and $b \to \infty$ independently?

404. Consider the function $f(x) = e^{ix^2}$. (a) Is $f(x)$ integrable on $(-\infty, \infty)$? (b) Has $f(x)$ an improper integral on $(-\infty, \infty)$?

405. Consider the improper integral $\int_{-\infty}^\infty f(x) \, dx$, where $f(x) = 0$ for $|x| < 1$, $f(x) = 1/x$ for $|x| \geqslant 1$. Make the substitution $x = y$ for $x < 0$, $x = 2y$ for $x \geqslant 0$, to get a new improper integral $\int_{-\infty}^\infty g(y) \, dy$. Find the values of the two improper integrals.

4.2 THE FOURIER INTEGRAL FORMULA: A HEURISTIC INTRODUCTION

We shall state and prove an important formula, formula (1) below, that will be the basis for much of our future work. For the validity of (1), the function $f(x)$ has to fulfill some conditions. We shall give such conditions in Definition 4.2.1, where we introduce an appropriate class of functions (denoted by DR after Dirichlet and the notation for the set of real numbers).

All known proofs of (1) are somewhat hard. We postpone a proof of (1) until Section 4.4, after some preliminary work is completed in Section 4.3. In this section (4.2) we introduce (1) by a heuristic argument. The basic idea in this argument is to replace $f(x)$ by a function that has the period $2a$ (where a is a given positive number) and that coincides with $f(x)$ on the interval $(-a, a)$, and then let $a \to \infty$.

4.2.1 Definition

We say that a complex-valued function $f(x)$ belongs to the *class DR* if it has the following four properties:

  1. $f(x)$ has the domain $(-\infty, \infty)$

2. $f(x)$ is integrable on $(-\infty, \infty)$
3. $f(x)$ is piecewise continuously differentiable on each finite interval
4. $f(x) = \frac{1}{2}[f(x^+) + f(x^-)]$ for each real number x.

4.2.2 The Fourier integral formula

Suppose that $f(x)$ is a function in the class DR, that x is a given real number, and that a is a positive number such that $|x| < a$. Then by Exercise 212c,

$$f(x) = \tfrac{1}{2}a_0 + \sum_{n=1}^{\infty} \left(a_n \cos \frac{n\pi x}{a} + b_n \sin \frac{n\pi x}{a} \right),$$

where

$$a_n = \frac{1}{a} \int_{-a}^{a} f(u) \cos \frac{n\pi u}{a}\, du, \qquad n \in N,$$

$$b_n = \frac{1}{a} \int_{-a}^{a} f(u) \sin \frac{n\pi u}{a}\, du, \qquad n \in Z^+.$$

It follows that

$$f(x) = \frac{1}{2a} \int_{-a}^{a} f(u)\, du + \sum_{n=1}^{\infty} \frac{1}{a} \int_{-a}^{a} f(u) \cos \frac{n\pi}{a}(x - u)\, du,$$

and further that

$$f(x) = \lim_{a \to \infty} \sum_{n=1}^{\infty} \frac{1}{a} \int_{-a}^{a} f(u) \cos \frac{n\pi}{a}(x - u)\, du.$$

It seems plausible that the following limit exists and is equal to $f(x)$:

(a) $$\lim_{a \to \infty} \sum_{n=1}^{\infty} \frac{1}{a} \int_{-\infty}^{\infty} f(u) \cos \frac{n\pi}{a}(x - u)\, du.$$

Let $\phi(t)$ denote the function

$$\phi(t) = \frac{1}{\pi} \int_{-\infty}^{\infty} f(u) \cos t(x - u)\, du.$$

The sum in (a) then is equal to

(b) $$\sum_{n=1}^{\infty} \frac{\pi}{a} \phi\left(\frac{n\pi}{a}\right).$$

This sum has an obvious geometric interpretation: If e.g. $\phi(t) > 0$ for $t > 0$, then the sum (b) is equal to the sum of the areas of certain rectangular regions that, for large values of a, approximate the region under the curve $y = \phi(t)$, $t > 0$ (see Figure 4.1). It is then plausible that $f(x)$ is equal to $\int_0^{\infty} \phi(t)\, dt$, or that

$$f(x) = \frac{1}{\pi} \int_0^{\infty} dt \int_{-\infty}^{\infty} f(u) \cos t(x - u)\, du. \tag{1}$$

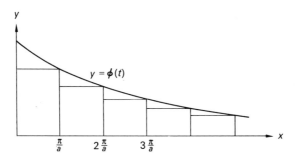

Figure 4.1

Formula (1) is called the *Fourier integral formula.* It holds true for $f(x)$ belonging to the class *DR*, but the above argument does not prove its truth. We want to present a rigorous proof for (1). To this end we shall need some auxiliary theorems.

EXERCISE

406. The Fourier integral formula can be transformed into the form (similar to formula (7) in Definition 1.2.5)

$$f(x) = \int_0^\infty [a(t) \cos tx + b(t) \sin tx] \, dt.$$

Find the functions $a(t)$ and $b(t)$ in terms of f, and compare with the formulas (6) in Definition 1.2.5 for the Fourier coefficients of a function.

4.3 AUXILIARY THEOREMS

4.3.1 Lemma. *Suppose that the function $f(x)$ is integrable on the infinite interval (α, ∞). Then*

$$\lim_{a \to \infty} \int_\alpha^\infty f(x) \sin ax \, dx = 0. \tag{1}$$

An analogous equality holds if $f(x)$ is integrable on $(-\infty, \alpha)$ or on $(-\infty, \infty)$.

Proof. Suppose that $f(x)$ satisfies the first condition and that $\varepsilon > 0$. There is a number β, $\beta > \alpha$, such that

$$\int_\beta^\infty |f(x)| \, dx < \frac{\varepsilon}{2}.$$

It follows that

$$\left| \int_\beta^\infty f(x) \sin ax \, dx \right| < \frac{\varepsilon}{2} \qquad \text{for} \quad a \in R.$$

By the Riemann–Lebesgue theorem (1.3.5) there is a real number a_0 such that

$$\left| \int_\alpha^\beta f(x) \sin ax \, dx \right| < \frac{\varepsilon}{2} \qquad \text{for} \quad a > a_0.$$

These two inequalities imply that

$$\left| \int_\alpha^\infty f(x) \sin ax \, dx \right| < \varepsilon \qquad \text{for} \quad a > a_0.$$

Now (1) follows. Proofs in the two remaining cases are constructed similarly.

4.3.2 Lemma. *Suppose that the function $f(x)$ belongs to the class DR. Then, for every real number x,*

$$f(x) = \lim_{a \to \infty} \frac{1}{\pi} \int_{-\infty}^{\infty} f(x + u) \frac{\sin au}{u} \, du. \tag{2}$$

Proof. Suppose that x is a given real number and that $\varepsilon > 0$. Then there is a number δ, $0 < \delta < \pi$, such that

(a) $$\frac{1}{\pi} \int_{-\delta}^{\delta} |f(x + u)| \, du < \varepsilon.$$

Associate with each value of a the integer n for which $n + \frac{1}{2} \leqslant a < n + \frac{3}{2}$. We shall show that the absolute value of the difference between any two successive integrals among

$$I_1 = \frac{1}{\pi} \int_{-\infty}^{\infty} f(x + u) \frac{\sin au}{u} \, du$$

$$I_2 = \frac{1}{\pi} \int_{-\delta}^{\delta} f(x + u) \frac{\sin au}{u} \, du$$

$$I_3 = \frac{1}{\pi} \int_{-\delta}^{\delta} f(x + u) \frac{\sin (n + \frac{1}{2})u}{u} \, du$$

$$I_4 = \frac{1}{\pi} \int_{-\delta}^{\delta} f(x + u) \frac{\sin (n + \frac{1}{2})u}{2 \sin \frac{1}{2}u} \, du$$

$$I_5 = \frac{1}{\pi} \int_{-\pi}^{\pi} f(x + u) \frac{\sin (n + \frac{1}{2})u}{2 \sin \frac{1}{2}u} \, du$$

is small, provided that a is large. The function $f(x + u)/u$ is integrable on the intervals $(-\infty, -\delta)$ and (δ, ∞). Lemma 4.3.1 then shows that there is a real number a_0 such that

$$|I_1 - I_2| < \varepsilon \qquad \text{for} \quad a > a_0.$$

It is seen that, for each real number u, $\left|\sin au - \sin(n + \tfrac{1}{2})u\right| \leqslant |u|$. Then the inequality (a) shows that

$$\left|I_2 - I_3\right| < \varepsilon \qquad \text{for} \quad a \in R.$$

The function $u^{-1} - (2\sin \tfrac{1}{2}u)^{-1}$ is bounded and integrable on the interval $(-\delta, \delta)$. The Riemann–Lebesgue theorem (1.3.5) then shows that there is a number $a_1 \geqslant a_0$ such that

$$\left|I_3 - I_4\right| < \varepsilon \qquad \text{for} \quad a > a_1.$$

The function $f(x + u)(2\sin \tfrac{1}{2}u)^{-1}$ is integrable on the intervals $(-\pi, -\delta)$ and (δ, π). One more application of the Riemann–Lebesgue theorem then shows the existence of a number $a_2 \geqslant a_1$ such that

$$\left|I_4 - I_5\right| < \varepsilon \qquad \text{for} \quad a > a_2.$$

Lemma 1.4.1 and Theorem 1.4.4 finally show that there is a number $a_3 \geqslant a_2$ such that

$$\left|I_5 - f(x)\right| < \varepsilon \qquad \text{for} \quad a > a_3.$$

Summing up we have

$$\left|I_1 - f(x)\right| < 5\varepsilon \qquad \text{for} \quad a > a_3.$$

From this the assertion follows by a standard argument.

In the statement of the following lemma it is assumed that both members of (b) are defined. It is left to the reader to prove this property for the left member (Exercise 409). For the right member this property is verified in the proof.

4.3.3 Lemma. *Suppose that the function $f(x)$ belongs to the class DR, that a is a given positive number, and that x is a given real number. Then*

(b)
$$\int_0^a dt \int_{-\infty}^\infty f(u) \cos t(x - u)\, du = \int_{-\infty}^\infty du \int_0^a f(u) \cos t(x - u)\, dt.$$

Proof. We first show that

(c)
$$\int_0^a dt \int_0^\infty f(u) \cos t(x - u)\, du = \int_0^\infty du \int_0^a f(u) \cos t(x - u)\, dt.$$

Denote by $u_0 = 0, u_1, \ldots, u_n, \ldots$ an unbounded and strictly increasing sequence of real numbers, such that the positive points of discontinuity of the function f, if any, belong to the sequence. A known theorem (see [13], p. 204) on interchange of the order of integration for a function of two variables, continuous on a closed rectangular domain, shows that for each $n \in Z^+$

$$\int_0^a dt \int_{u_{n-1}}^{u_n} f(u) \cos t(x - u)\, du = \int_{u_{n-1}}^{u_n} du \int_0^a f(u) \cos t(x - u)\, dt.$$

It is then seen that, for each positive number y,

(d) $\qquad \int_0^a dt \int_0^y f(u) \cos t(x - u)\, du = \int_0^y du \int_0^a f(u) \cos t(x - u)\, dt.$

Suppose that $\varepsilon > 0$. There is a positive number y_0 such that

$$\int_{y_0}^\infty |f(u)|\, du < \frac{\varepsilon}{a}.$$

The absolute value of the difference of the left members of (c) and (d) then is less than ε for $y > y_0$. Hence the left member of (c) is the limit of the right member of (d) as $y \to \infty$. The right member of (c) is this limit, by definition. This proves (c). An equation analogous to (c), but with the negative u-axis as interval of integration, is proved in the same way. Addition member by member of (c) to this equation gives (b).

EXERCISES

*407. State and prove a corollary to Lemma 4.3.2, analogous to Corollary 1.4.5, for functions having properties 1, 2, 3 in Definition 4.2.1.

408. Is it legitimate to replace the words "is integrable" by "has an improper integral" in the statement of Lemma 4.3.1?

409. Show that the interior integral in the left member of (c) is a continuous function of t.
Hint. Denote the integrand by $\phi(u, t)$. First show that the integral $\int_0^\infty \phi(u, t)\, du$ is *uniformly convergent* on the closed interval $[0, a]$, i.e. that there corresponds to every number $\varepsilon > 0$ a number y_0 such that $\left| \int_y^\infty \phi(u, t)\, du \right| < \varepsilon/3$ for $y > y_0$ and every $t \in [0, a]$. Then estimate the expression $\left| \int_0^\infty \phi(u, t + h)\, du - \int_0^\infty \phi(u, t)\, du \right|$ for small values of $|h|$.

4.4 PROOF OF THE FOURIER INTEGRAL FORMULA. FOURIER TRANSFORMS

4.4.1 Theorem (giving the Fourier integral formula). *Suppose that the function $f(x)$ belongs to the class DR. Then, for every real number x,*

$$f(x) = \frac{1}{\pi} \int_0^\infty dt \int_{-\infty}^\infty f(u) \cos t(x - u)\, du. \tag{1}$$

Proof. We have by Lemma 4.3.3 for $a > 0$

$$\frac{1}{\pi} \int_0^a dt \int_{-\infty}^\infty f(u) \cos t(x - u)\, du = \frac{1}{\pi} \int_{-\infty}^\infty du \int_0^a f(u) \cos t(x - u)\, dt$$

$$= \frac{1}{\pi} \int_{-\infty}^\infty f(u) \frac{\sin a(x - u)}{x - u}\, du = \frac{1}{\pi} \int_{-\infty}^\infty f(x + u) \frac{\sin au}{u}\, du.$$

By Lemma 4.3.2, the last member tends to $f(x)$ as $a \to \infty$. Then also the first member tends to $f(x)$. This is the assertion.

In the following three corollaries we introduce three functions, denoted by $F(t)$, $F_c(t)$, $F_s(t)$. The domain of the function $F(t)$ is $(-\infty, \infty)$. For the two remaining functions we take $[0, \infty)$ as domain.

4.4.2 Corollary. *Suppose that the function $f(x)$ belongs to the class DR. Set*

$$F(t) = \frac{1}{2\pi} \int_{-\infty}^{\infty} f(x) e^{-itx}\, dx. \tag{2}$$

Then, for all real numbers x,

$$f(x) = \int_{-\infty}^{\infty} F(t) e^{itx}\, dt. \tag{3}$$

Proof. The interior integral in formula (1) is an even function of t. The integral

$$\int_{-\infty}^{\infty} f(u) \sin t(x - u)\, du$$

is an odd function of t. Hence this function has the integral zero on every interval $(-a, a)$. Then (1) shows that, for every real number x,

$$f(x) = \frac{1}{2\pi} \int_{-\infty}^{\infty} dt \int_{-\infty}^{\infty} f(u) e^{it(x-u)}\, du$$

$$= \int_{-\infty}^{\infty} \left[\frac{1}{2\pi} \int_{-\infty}^{\infty} f(u) e^{-itu}\, du \right] e^{itx}\, dt.$$

This is the assertion.

4.4.3 Definitions

The function $F(t)$, defined by (2), is called the *Fourier transform* of $f(x)$. Further, with regard to (3), $f(x)$ is called the *inverse Fourier transform* of $F(t)$.

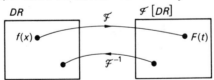

Figure 4.2

We let $\mathscr{F}$ denote the operator that maps every function $f(x)$ of the class DR on its Fourier transform $F(t)$. We shall denote this by $\mathscr{F}[f] = F$ or, more detailed, $\mathscr{F}[f(x)](t) = F(t)$. In Figure 4.2 we let the points inside the left rectangle represent the functions in the class DR; analogously for the right rectangle. Corollary 4.4.2 shows that the operator $\mathscr{F}$ is one-to-one. Then it has an inverse operator $\mathscr{F}^{-1}$, such that $\mathscr{F}^{-1}[F] = f$ or $\mathscr{F}^{-1}[F(t)](x) = f(x)$. A figure similar to Figure 4.2 can be drawn for every transform that we shall introduce in this text.

It should be observed that the hypotheses in the following Corollary 4.4.4 imply that the function $f(x)$ is continuous from the right at the point $x = 0$, and in Corollary 4.4.6 that $f(0) = 0$.

4.4.4 Corollary. *Suppose that the function $f(x)$ has the domain $[0, \infty)$, and that $f(x)$ has an extension that is an even function on $(-\infty, \infty)$ and belongs to the class DR. Set for $t \geqslant 0$*

$$F_c(t) = \frac{2}{\pi} \int_0^\infty f(x) \cos tx \, dx. \tag{4}$$

Then, for all nonnegative real numbers x,

$$f(x) = \int_0^\infty F_c(t) \cos tx \, dt. \tag{5}$$

Proof. Denote the even extension of $f(x)$ by $f_e(x)$. We have by (1) for every real number x

$$f_e(x) = \frac{1}{\pi} \int_0^\infty dt \int_{-\infty}^\infty f_e(u) \left[\cos tx \cos tu + \sin tx \sin tu \right] du$$

$$= \frac{2}{\pi} \int_0^\infty dt \int_0^\infty f(u) \cos tx \cos tu \, du.$$

There follows for $x \geqslant 0$

$$f(x) = \int_0^\infty \left[\frac{2}{\pi} \int_0^\infty f(u) \cos tu \, du \right] \cos tx \, dt.$$

This is the assertion.

4.4.5 Definitions

The function $F_c(t)$, defined by (4), is called the *cosine transform* of $f(x)$. Further, with regard to (5), $f(x)$ is called the *inverse cosine transform* of $F_c(t)$.

4.4.6 Corollary. *Suppose that the function $f(x)$ has the domain $[0, \infty)$ and that $f(x)$ has an extension that is an odd function on $(-\infty, \infty)$ and belongs to the class DR. Set for $t \geqslant 0$*

$$F_s(t) = \frac{2}{\pi} \int_0^\infty f(x) \sin tx \, dx. \tag{6}$$

Then, for all nonnegative real numbers x,

$$f(x) = \int_0^\infty F_s(t) \sin tx \, dt. \tag{7}$$

Proof. Denote the odd extension of $f(x)$ by $f_o(x)$. We have by (1) for every real number x

$$f_o(x) = \frac{1}{\pi} \int_0^\infty dt \int_{-\infty}^\infty f_o(u) \left[\cos tx \cos tu + \sin tx \sin tu \right] du$$

$$= \frac{2}{\pi} \int_0^\infty dt \int_0^\infty f(u) \sin tx \sin tu \, du.$$

There follows for $x \geqslant 0$

$$f(x) = \int_0^\infty \left[\frac{2}{\pi} \int_0^\infty f(u) \sin tu \, du \right] \sin tx \, dt.$$

This is the assertion.

4.4.7 Definitions

The function $F_s(t)$, defined by (6), is called the *sine transform* of $f(x)$. Further, with regard to (7), $f(x)$ is called the *inverse sine transform* of $F_s(t)$.

4.4.8 Definition

The formulas (3), (5) and (7) are called the *inversion formulas* for the Fourier transform, the cosine transform and the sine transform respectively.

4.4.9 Definitions

Suppose that the function $f(x)$ has the domain $(-\infty, \infty)$ with the possible exception of finitely many points in each finite interval, and that there exists a function $f_1(x)$ in the class DR, such that $f(x) = f_1(x)$ except perhaps for finitely many points in each finite interval. Then the function $F(t)$ which is defined by (2) is called the *Fourier transform* of $f(x)$. A valid equation is obtained if the left member of (3) is replaced by $f_1(x)$. The notions of a *sine transform* and a *cosine transform* are extended analogously, and the equations (5) and (7) are adjusted analogously; it is left to the reader to formulate these extensions and adjustments.

The integrals in formulas (1) through (7) are called *Fourier integrals*. There is no universal agreement among authors on Fourier integrals how to phrase formulas (2) through (7). Some authors place the factor $1/\sqrt{2\pi}$ in front of the integrals in (2) and (3) and the factor $\sqrt{2/\pi}$ in front of the integrals in (4) through (7) in order to make the expressions more symmetric. Some authors place a minus sign in the exponent in (3), not in (2). Still more variants of the formulas could be cited. Formulas (2) through (7) have some mnemonic merits: The formulas (2) and (3) remind one of formulas (16) and (17) respectively in Example 2.2.7, formulas (4) through (7) remind one of formulas (10), (11), (13), (14) respectively in Examples 2.2.5 and 2.2.6.

4.4.10 Example

a) Suppose that $a > 0$. Find the cosine transform $F_c(t)$ and the sine transform $F_s(t)$ of the function $f(x) = e^{-ax}$, $x > 0$.
b) Write down the corresponding inversion formulas.

Solution. a) Formulas (4) and (6) give

$$F_c(t) = \frac{2}{\pi} \int_0^\infty e^{-ax} \cos tx \, dx, \qquad F_s(t) = \frac{2}{\pi} \int_0^\infty e^{-ax} \sin tx \, dx,$$

$$F_c(t) + iF_s(t) = \frac{2}{\pi}\int_0^\infty e^{(-a+it)x}\,dx = \frac{2}{\pi}\frac{1}{a-it} = \frac{2}{\pi}\frac{a+it}{a^2+t^2},$$

$$F_c(t) = \frac{2}{\pi}\frac{a}{a^2+t^2}, \qquad F_s(t) = \frac{2}{\pi}\frac{t}{a^2+t^2}.$$

b) Formulas (5) and (7) give

$$\frac{2}{\pi}\int_0^\infty \frac{a}{a^2+t^2}\cos tx\,dt = e^{-ax} \qquad \text{for } x \geqslant 0,$$

$$\frac{2}{\pi}\int_0^\infty \frac{t}{a^2+t^2}\sin tx\,dt = \begin{cases} e^{-ax} & \text{for } x > 0 \\ 0 & \text{for } x = 0. \end{cases}$$

4.4.11 Example

a) Suppose that $a > 0$. Find the cosine transform $F_c(t)$ of the function $f(x)$, defined by $f(x) = 1$ for $0 < x < a$, $f(x) = 0$ for $a < x$.

b) Write down the corresponding inversion formula.

c) Use the result to compute the integral $\displaystyle\int_0^\infty \frac{\sin t}{t}\,dt$.

Solution.   a) Formula (4) gives

$$F_c(t) = \frac{2}{\pi}\int_0^a \cos tx\,dx = \frac{2}{\pi}\frac{\sin at}{t} \qquad \text{for } t > 0, \qquad F_c(0) = \frac{2a}{\pi}.$$

b) Formula (5) gives

$$\frac{2}{\pi}\int_0^\infty \frac{\sin at}{t}\cos tx\,dt = \begin{cases} 1 & \text{for } 0 \leqslant x < a \\ \frac{1}{2} & \text{for } x = a \\ 0 & \text{for } a < x. \end{cases}$$

c) This result gives for $x = 0$ and $a = 1$ (or $x = a = \frac{1}{2}$)

$$\int_0^\infty \frac{\sin t}{t}\,dt = \frac{\pi}{2}. \tag{8}$$

4.4.12 Example

Find the Fourier transform $F(t)$ of the function $f(x) = e^{-x^2/2}$.

Solution. Formula (2) gives (differentiation under the integral sign is legitimate; see Exercise 414)

$$F(t) = \frac{1}{2\pi}\int_{-\infty}^\infty e^{-x^2/2}e^{-itx}\,dx, \qquad F'(t) = \frac{-i}{2\pi}\int_{-\infty}^\infty xe^{-x^2/2}e^{-itx}\,dx$$

$$= \frac{i}{2\pi}\left[e^{-x^2/2}e^{-itx}\right]_{x=-\infty}^{x=\infty} - \frac{t}{2\pi}\int_{-\infty}^\infty e^{-x^2/2}e^{-itx}\,dx.$$

Hence $F'(t) = -tF(t)$. Thus (cf. Lemma 5.4.2) there is a number c such that $F(t) = ce^{-t^2/2}$. The substitution $x = t/\sqrt{2}$ applied to the formula (see Exercise 414)

$$\int_{-\infty}^{\infty} e^{-x^2}\, dx = \sqrt{\pi} \tag{9}$$

shows that $c = F(0) = \dfrac{1}{2\pi}\sqrt{2\pi} = 1/\sqrt{2\pi}$. Hence

$$F(t) = \frac{1}{\sqrt{2\pi}}\, e^{-t^2/2}.$$

The next example demonstrates that certain Fourier transforms can be computed by use of complex analysis (see [1], especially the residue theorem pp. 147–151). We let z denote a complex variable. We use the notation $\operatorname{Res}_{z=a} f(z)$ for the residue of the function $f(z)$ at the point $z = a$. We use the fact that if the functions $g(z)$ and $h(z)$ are analytic in a neighborhood of the point $z = a$, if $g(a) \neq 0$, and if $h(z)$ has a simple zero at the point $z = a$, then $\operatorname{Res}_{z=a} g(z)/h(z) = g(a)/h'(a)$. We need a lemma on integrals along certain half-circles in the complex plane (see Figure 4.3) and a lemma on certain integrals on $(-\infty, \infty)$.

4.4.13 Lemma. *Suppose that the function $f(z)$ is analytic for $|z| > a$. Let C_r denote the half-circle $z = re^{i\theta}$, $0 \leqslant \theta \leqslant \pi$. Suppose that $t > 0$, and that $f(z) \to 0$ as $z \to \infty$. Then*

$$\int_{C_r} f(z)\, e^{itz}\, dz \to 0 \qquad as \quad r \to \infty. \tag{10}$$

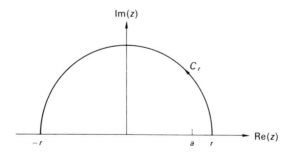

Figure 4.3

Proof. Suppose that $\varepsilon > 0$. There is a positive number r_0 such that, for $r > r_0$, $|f(z)| < \varepsilon$ on C_r. As $\sin\theta > 2\theta/\pi$ for $0 < \theta < \pi/2$, we have for $r > r_0$:

$$\left|\int_{C_r} f(z)e^{itz} dz\right| = \left|\int_0^\pi f(re^{i\theta})e^{itr(\cos\theta + i\sin\theta)}ire^{i\theta} d\theta\right|$$

$$< \varepsilon r\int_0^\pi e^{-tr\sin\theta} d\theta = 2\varepsilon r\int_0^{\pi/2} e^{-tr\sin\theta} d\theta > 2\varepsilon r\int_0^{\pi/2} e^{-2tr\theta/\pi} d\theta$$

$$= 2\varepsilon r\frac{\pi}{2tr}[e^{-2tr\theta/\pi}]_{\theta=\pi/2}^{\theta=0} < \frac{\varepsilon\pi}{t}.$$

The assertion follows.

4.4.14 Lemma. *Suppose that $t > 0$. Suppose that $R(z)$ is a rational function with real coefficients and no poles on the real axis, and that $R(z) \to 0$ as $z \to \infty$. Suppose that the residues at the poles of the function $R(z)e^{itz}$ in the half-plane $\text{Im}(z) > 0$ are $r_1, r_2, \ldots, r_n$. Then*

$$\int_{-\infty}^\infty R(x)e^{itx} dx = 2\pi i \sum_{\nu=1}^n r_\nu. \qquad (11)$$

Proof. Suppose that r is a positive number, so large that the poles are all situated inside the circle $|z| = r$. Integration around the half-circle of Fig. 4.3 gives, by the residue theorem,

$$\int_{-r}^r R(x)e^{itx} dx + \int_{C_r} R(z)e^{itz} dz = 2\pi i\sum_{\nu=1}^n r_\nu.$$

Here, let $r \to \infty$, and apply Lemma 4.4.13. Then (11) follows.

4.4.15 Example

Find the Fourier transform $F(t)$ of the function

$$f(x) = \frac{2}{x^2 - 2x + 2}.$$

Solution. **1.** Suppose that $t < 0$. Lemma 4.4.14 shows that

$$F(t) = \frac{1}{2\pi}\int_{-\infty}^\infty f(x)e^{-itx} dx = \frac{1}{2\pi}2\pi i \operatorname*{Res}_{z=1+i} \frac{2}{z^2 - 2z + 2} e^{-itz}$$

$$= i\left[\frac{2e^{-itz}}{2z - 2}\right]_{z=1+i} = i\frac{e^{-it(1+i)}}{i} = e^t(\cos t - i\sin t).$$

2. Suppose that $t = 0$. The result in part **1** and the first result in Exercise 419 show that $F(0) = 1$.

3. Suppose that $t > 0$. The result in part **1** shows that
$$F(-t) = e^{-t}(\cos t + i \sin t).$$

Since the function $f(x)$ is real-valued, it is readily verified that $F(t) = \overline{F(-t)}$. Hence
$$F(t) = e^{-t}(\cos t - i \sin t).$$

4. Summing up, we have for all real values of t
$$F(t) = e^{-|t|} (\cos t - i \sin t).$$

Exercises 417–421 exhibit some important properties of Fourier transforms. Certain among these properties have counterparts for the sine and cosine transforms. (For some of them this can be proved by means of the formulas in Exercise 416.) The second claim in Exercise 419 is the *Riemann–Lebesgue theorem* for Fourier transforms. In Exercises 420 and 421 it is shown that, under certain hypotheses, differentiation and integration of a function correspond to certain simple operations performed on its Fourier transform.

EXERCISES

410. a) Find the Fourier transform of the function
$$f(x) = \begin{cases} 1 - |x| & \text{for} \quad |x| \leq 1 \\ 0 & \text{for} \quad |x| > 1. \end{cases}$$
   b) Write down the corresponding inversion formula.

411. a) Find the cosine transform of the function
$$f(x) = \begin{cases} \cos x & \text{for} \quad 0 \leq x \leq \pi/2 \\ 0 & \text{for} \quad \pi/2 < x. \end{cases}$$
   b) Use the result of (a) to compute the integral
$$\int_0^\infty \frac{\cos (\pi t/2)}{1 - t^2} \, dt.$$

412. a) Find the sine transform of the function
$$f(x) = e^{-x} \sin x, \quad x > 0.$$
   b) Use the result of (a) to compute the integral
$$\int_0^\infty \frac{4t}{t^4 + 4} \sin t \, dt.$$

*413. By use of the method in Example 4.4.15, compute the Fourier transforms of the functions

a) $\dfrac{1}{1 + x^2}$,

b) $\dfrac{1}{(1 + x^2)^2}$,

c) $\dfrac{1}{(x^2 + 2x + 2)(x^2 + 2x + 5)}$.

414. Complete the solution of Example 4.4.12.
     Hints. For the differentiation under the integral sign, use the technique in the proof
     of Theorem 5.1.4. For a proof of formula (9), observe that transformation to polar
     coordinates r, θ gives (cf. [13], Theorem 9.32):

$$\left[\int_{-\infty}^{\infty} e^{-x^2}\, dx\right]^2 = \int_{-\infty}^{\infty} e^{-x^2}\, dx \int_{-\infty}^{\infty} e^{-y^2}\, dy = \int_{0}^{2\pi}\int_{0}^{\infty} e^{-r^2}\, r\, dr\, d\theta = \pi.$$

415. Suppose that the function $f(x)$ belongs to the class DR. Does it follow that its Fourier
     transform belongs to the same class?
     Hint. Set $f(x) = e^{-x}$ for $x > 0$, $f(0) = \frac{1}{2}$, $f(x) = 0$ for $x < 0$, and study the Fourier
     transform of $f(x)$. The result should be compared with the comments to Definition
     4.1.1.

416. a) Make the same assumptions as in Corollary 4.4.4. Further let $F(t)$ denote the
     Fourier transform of the even extension of $f(x)$ considered in the corollary. Show
     that, for $t \geqslant 0$,

$$F_c(t) = 2F(t).$$

   b) Under analogous assumptions show that, for $t \geqslant 0$,

$$F_s(t) = 2iF(t).$$

417. Suppose that the functions $f(x)$ and $g(x)$ belong to the class DR, and that their Fourier
     transforms are $F(t)$ and $G(t)$ respectively. Suppose that α and β are complex numbers,
     and that a is a nonzero real number. Find the Fourier transforms of the functions
     a) $\alpha f(x) + \beta g(x)$,   b) $e^{iax}f(x)$,   c) $f(ax)$.

418. Suppose that the functions $f(x)$ and $xf(x)$ both belong to the class DR, and that
     $f(x)$ has the Fourier transform $F(t)$. Show that $xf(x)$ has the Fourier transform $iF'(t)$.

419. Suppose that the function $f(x)$ belongs to the class DR, and that $f(x)$ has the Fourier
     transform $F(t)$. Show that $F(t)$ is a continuous function, and that $F(t) \to 0$ as $|t| \to \infty$.
     Hints. For proof of the first assertion apply the method used in Exercise 409; for
     the second assertion use the method in the proof of Lemma 4.3.1.

420. Suppose that the function $f(x)$ and its derivative $f'(x)$ both belong to the class DR.
     Also suppose that $f(x)$ tends to zero as $|x| \to \infty$. Denote the Fourier transform of
     $f(x)$ by $F(t)$. Show that $f'(x)$ has the Fourier transform $itF(t)$.
     Hint. Use partial integration of the expression for the Fourier transform of $f'(x)$.

421. Suppose that the function $f(x)$ and its primitive function $p(x) = \int_{0}^{x} f(u)\,du$ both belong to
     the class DR. Also suppose that $p(x)$ tends to zero as $|x| \to \infty$. Denote the Fourier
     transform of $f(x)$ by $F(t)$. Show that $p(x)$ has the Fourier transform $(1/it)F(t)$ for $t \neq 0$ and
     $\lim_{t\to 0} (1/it)F(t)$ for $t = 0$.
     Hint. Apply Exercise 420 and the first claim in Exercise 419.

4.5 THE CONVOLUTION THEOREM. THE PARSEVAL FORMULA

Theorem 4.5.5 below, which is called the convolution theorem for Fourier transforms, and
formula (5) below, which is called the Parseval formula for Fourier transforms and should

be compared with formula (5) in Example 2.3.7, are best discussed for Lebesgue-integrable functions (the assumptions can then be chosen considerably wider than below; cf. [4], pp. 20 and 48). Here they are given only for a fairly restricted class of functions, denoted by D_0 (here D is for Dirichlet; the subscript zero denotes a bounded support). For several integrals we shall use $-\infty$ and ∞ as limits of integration in contexts where finite limits could be used; we do so partly for practical reasons, partly to exhibit the same formulas as in a more general theory.

4.5.1 Definitions
Given a function $f(x)$, we shall call the set $\{x; x \in D_f$ and $f(x) \neq 0\}$ its *support*. We say that a function $f(x)$ belongs to the *class* D_0 if it belongs to the class DR and its support is bounded.

4.5.2 Comments
Hence each function in the class D_0 vanishes outside some finite interval $[-a, a]$. For example the functions $f_n(x) = \max(n^2 - x^2, 0)$, $n \in N$, all belong to the class D_0. As the class D_0 is a subclass of the class DR, the inversion formula for Fourier transforms, formula (3) in Corollary 4.4.2, holds for every function in the class D_0.

4.5.3 Definition
Suppose that the functions $f(x)$ and $g(x)$ belong to the class D_0. Their *convolution*, denoted by $f * g$, is the function defined by

$$(f * g)(x) = \int_{-\infty}^{\infty} f(y)g(x - y)\,dy. \tag{1}$$

4.5.4 Comments
Suppose that a is a positive number such that $f(x)$ and $g(x)$ both vanish for $|x| > a$. Then the integrand in (1) is zero outside the shaded domain Ω in Fig. 4.4a. The

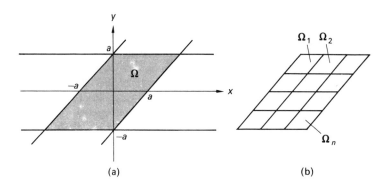

(a)   (b)

Figure 4.4

assumption that $f(x)$ and $g(x)$ belong to the class D_0 implies that Ω can be divided into a finite number of domains Ω_ν (see Fig. 4.4b) by lines parallel to the sides of Ω such that, in the interior of each Ω_ν, the integrand is equal to a function that is continuous on, say, the whole plane. It follows that the substitutions and the changes of the order of integration which we carry out in the following are legitimate.

The substitution $x - y = u$ gives

$$(f * g)\,(x) = \int_{-\infty}^{\infty} f(y)g(x - y)\,dy = \int_{-\infty}^{\infty} g(u)f(x - u)\,du = (g * f)\,(x).$$

Hence the operation to form the convolution of two functions is commutative:

$$f * g = g * f. \tag{2}$$

Theorem 4.5.5 states that the Fourier transform of the convolution of two functions is 2π times the product of the Fourier transforms of the two functions. Corollary 4.4.2 gives a corollary for Theorem 4.5.5.

4.5.5 Theorem (the convolution theorem for Fourier transforms). *Suppose that the functions $f(x)$ and $g(x)$ belong to the class D_0. Then their convolution $(f*g)(x)$ belongs to the class D_0, and the Fourier transforms satisfy the equation*

$$\mathscr{F}[f * g]\,(t) = 2\pi\mathscr{F}[f]\,(t)\mathscr{F}[g]\,(t). \tag{3}$$

Proof. **1.** Set $h(x) = (f * g)(x)$. Obviously $h(x)$ is a continuous function with domain $(-\infty, \infty)$; further it vanishes outside a certain finite interval. Hence $h(x)$ has all the properties of a function in the class D_0 except perhaps property 3 of Definition 4.2.1. A straight line $x = a$ intersects the boundaries of the regions Ω_ν in Figure 4.4b in finitely many points $y_0(a) < y_1(a) < \cdots < y_m(a)$; here m depends on a. The projections of the vertices of the domains Ω_ν on the x-axis are finitely many points $x_0 < x_1 < \cdots < x_k$. In each of the closed intervals $[x_{\varkappa - 1}, x_\varkappa], \varkappa \in \{1, 2, \ldots, k\}$, $h(x)$ has a representation of the form

$$h(x) = \sum_{\mu=1}^{m} \int_{y_{\mu-1}(x)}^{y_\mu(x)} f(y)g(x - y)\,dy,$$

where m depends on $\varkappa$, where one or more among the intervals of integration degenerate to points for $x = x_{\varkappa-1}$ and for $x = x_\varkappa$, and where, for each nondegenerate interval of integration, it is true that the integrand and its partial derivative with respect to x in the open interval are equal to functions that are continuous on the closed interval. Thus on the interval $[x_{\varkappa-1}, x_\varkappa]$ the function $h(x)$ is equal to a function that has a continuous derivative on the interval. It follows that $h(x)$ has property 3 of Definition 4.2.1. Hence $h(x)$ belongs to the class D_0.

2. Change of the order of integration and the substitution $x - y = u$ give

$$\mathscr{F}[f * g]\,(t) = \frac{1}{2\pi} \int_{-\infty}^{\infty} (f * g)(x)\,e^{-itx}\,dx$$

$$= \frac{1}{2\pi} \int_{-\infty}^{\infty} e^{-itx}\,dx \int_{-\infty}^{\infty} f(y)g(x-y)\,dy$$

$$= \frac{1}{2\pi} \int_{-\infty}^{\infty} f(y)\,dy \int_{-\infty}^{\infty} g(x-y)\,e^{-itx}\,dx$$

$$= 2\pi \frac{1}{2\pi} \int_{-\infty}^{\infty} f(y)\,e^{-ity}\,dy \frac{1}{2\pi} \int_{-\infty}^{\infty} g(u)\,e^{-itu}\,du$$

$$= 2\pi \mathscr{F}[f](t)\mathscr{F}[g](t).$$

This proves formula (3).

The above proof shows among other things that the convolution $h(x)$ is continuous and piecewise differentiable. It need not, however, be a differentiable function (cf. Exercise 422).

4.5.6 Corollary. *Suppose that the functions $f(x)$ and $g(x)$ belong to the class D_0 and have the Fourier transforms $F(t)$ and $G(t)$ respectively. Then*

$$(f*g)(x) = 2\pi \int_{-\infty}^{\infty} F(t)G(t)\,e^{itx}\,dt. \tag{4}$$

4.5.7 Theorem (giving the Parseval formula for Fourier transforms). *Suppose that the function $f(x)$ belongs to the class D_0. Then*

$$\int_{-\infty}^{\infty} |F(t)|^2\,dt = \frac{1}{2\pi} \int_{-\infty}^{\infty} |f(x)|^2\,dx. \tag{5}$$

Proof. In Corollary 4.5.6, set $g(x) = \overline{f(-x)}$. Then

$$G(t) = \frac{1}{2\pi} \int_{-\infty}^{\infty} \overline{f(-x)}\,e^{-itx}\,dx = \frac{1}{2\pi} \int_{-\infty}^{\infty} \overline{f(y)}\,e^{ity}\,dy = \overline{F(t)}.$$

Formula (4) now yields

$$\int_{-\infty}^{\infty} f(y)\overline{f(y-x)}\,dy = 2\pi \int_{-\infty}^{\infty} F(t)\,\overline{F(t)}\,e^{itx}\,dt.$$

Here set $x = 0$, and (5) is obtained.

EXERCISES

422. Find the Fourier transform of the function $f(x)$ defined by $f(x) = 1$ for $|x| < 1$, $f(x) = 0$ for $|x| > 1$. Then find the inverse Fourier transform of the function $(2/\pi)\,[(\sin t)/t]^2$.

423. If Theorem 4.5.7 is applied to the function $f(x)$, defined by $f(x) = x$ for $|x| < 1$, $f(x) = 0$ for $|x| > 1$, a certain equation is obtained. Which equation?

*424. State and prove analogs to Theorem 4.5.7 for the cosine and sine transforms. *Hint.* Apply the results in Exercise 416.

CHAPTER 5

LAPLACE TRANSFORMS

5.1 DEFINITION OF THE LAPLACE TRANSFORM. DOMAIN. ANALYTICITY

Certain functions in this chapter are complex-valued functions of a complex variable. The reader is assumed to have some basic knowledge of complex analysis. In particular use will be made of the concept of a function analytic in a domain, of the concept of an integral in the complex plane, of the residue theorem, and of some properties of power series (see [1], the sections that begin on pp. 69, 109, 147 and 177).

We shall associate with every function $f(t)$ that satisfies certain conditions a function, called the Laplace transform of $f(t)$, and denoted by $F(s)$ or $\mathscr{L}[f](s)$ or $\mathscr{L}[f(t)]$ (s). To be able to give an inversion formula for this transform, we introduce a class of functions, denoted by E since an exponential function appears in the definition of the class.

We shall use t and $s = \sigma + i\omega$ to denote the independent variable of a function in the class E and of its Laplace transform respectively; in practical applications t generally denotes time and ω denotes an angular velocity.

5.1.1 Definitions

We say that a complex-valued function $f(t)$ belongs to the *class E*, if its domain is $(-\infty, \infty)$, if $f(t) = 0$ for $t < 0$, and if there is a real number α such that the function $f(t)e^{-\alpha t}$ belongs to the class DR of Definition 4.2.1.

Suppose that the function $f(t)$ belongs to the class E, or that the restriction of $f(t)$ to $(0, \infty)$ is equal to the restriction to the same interval of a function $f_1(t)$ in the class E except perhaps at finitely many points in each finite interval. By the *Laplace transform* of $f(t)$ we mean the function

$$\mathscr{L}[f](s) = F(s) = \int_0^{\infty} f(t)e^{-st}\,dt, \tag{1}$$

where $s = \sigma + i\omega$ is a complex variable, and where we let the domain of $F(s)$ be the set of all complex numbers s such that the integrand in (1) is integrable on $(0, \infty)$.

Some authors on Laplace transforms only require that the integrand in (1) have an improper integral on $(0, \infty)$. (The domain of $F(s)$ may then be larger.) By introducing the definition above, however, we get a theory that is simpler and probably equally useful.

The set of all complex numbers is denoted by C.

5.1.2 Example

Suppose that $a \in C$. Find the Laplace transform of the function $f(t) = e^{at}$.

Solution. Formula (1) gives

$$F(s) = \int_0^\infty e^{at} e^{-st}\, dt = \int_0^\infty e^{-(s-a)t}\, dt.$$

Here the integrand is integrable for $\mathrm{Re}(s) > \mathrm{Re}(a)$. Hence

(a) $\qquad F(s) = \left[\dfrac{e^{-(s-a)t}}{-(s-a)}\right]_{t=0}^{t=\infty} = \dfrac{1}{s-a}$ for $\mathrm{Re}(s) > \mathrm{Re}(a)$.

Formula (a) answers the question in Example 5.1.2 in accordance with Definition 5.1.1. In the following, however, we do not specify the domain of a Laplace transform as in (a), nor do we specify the domain of validity of an equation containing two or more Laplace transforms (an example is formula (1) in Theorem 5.3.1). We then have the formula

$$\mathscr{L}[e^{at}](s) = \frac{1}{s-a} \qquad \text{for} \quad a \in C. \tag{2}$$

The Laplace transform of e^{at} is a function of s that is analytic in a half-plane, viz. the half-plane $\mathrm{Re}(s) > \mathrm{Re}(a)$. This is a special case of the following lemma and theorem.

5.1.3 Lemma. *Suppose that the function $f(t)$ belongs to the class E. Then either the domain of its Laplace transform $F(s)$ is the whole s-plane, or there exists a real number γ such that the domain is the open half-plane $\mathrm{Re}(s) > \gamma$ or the closed half-plane $\mathrm{Re}(s) \geqslant \gamma$.*

Proof. By the definition of the class E there is a real number α, and also a complex number a, such that the function $f(t)e^{-at}$ is integrable on $(0, \infty)$. Let s denote a complex number such that $\mathrm{Re}(s) \geqslant \mathrm{Re}(a)$. Then

$$\int_0^x \left|f(t)e^{-st}\right| dt \leqslant \int_0^x \left|f(t)e^{-at}\right| dt \leqslant \int_0^\infty \left|f(t)e^{-at}\right| dt.$$

It follows that the left member tends to a limit as $x \to \infty$, that $f(t)e^{-st}$ is integrable on $(0, \infty)$, that s belongs to the domain D_F, and that D_F has the half-plane $\mathrm{Re}(s) \geqslant \mathrm{Re}(a)$ as a subset. Now consider the union of all half-planes corresponding to all complex numbers a with the mentioned property. This union is the domain of $F(s)$, and it is either the whole s-plane or one of the half-planes in the statement of the lemma. This proves the assertion.

In the following theorem it is proved that the function $F(s)$ is analytic in the interior of its domain, i.e. in the whole s-plane in the first case of the lemma, and in the open half-plane $\mathrm{Re}(s) > \gamma$ in the two remaining cases. It is also proved that the derivative of $F(s)$ can be obtained by differentiation under the integral sign.

5.1.4 Theorem. *Suppose that the function $f(t)$ belongs to the class E. Denote the Laplace transform of $f(t)$ by $F(s)$, and denote the interior of the domain of $F(s)$ by Ω. Then the function $F(s)$ is analytic in Ω, and its derivative is given by the formula:*

$$F'(s) = -\int_0^\infty tf(t)e^{-st}\, dt \qquad \text{for} \quad s \in \Omega. \tag{3}$$

Proof. Suppose that $s = \sigma + i\omega \in \Omega$, that δ is a positive number, so small that the circle with center s and radius δ is situated in Ω, and that $h, 0 < |h| < \delta$, is a complex number. Then it follows that (it is readily verified that the integrands are integrable on $(0, \infty)$):

$$\left| \frac{F(s+h) - F(s)}{h} + \int_0^\infty tf(t)e^{-st}\, dt \right|$$

$$= \left| \int_0^\infty \left[\frac{e^{-ht} - 1}{h} + t \right] f(t)e^{-st}\, dt \right| \leqslant |h| \int_0^\infty t^2 |f(t)| e^{-(\sigma - \delta)t}\, dt,$$

because, for $h \neq 0$ and $t > 0$,

$$\left| \frac{e^{-ht} - 1}{h} + t \right| = \left| \frac{ht^2}{2!} - \frac{h^2 t^3}{3!} + \cdots + (-1)^n \frac{h^{n+1} t^n}{n!} + \cdots \right|$$

$$< |h| t^2 \left(1 + \frac{|h|t}{1!} + \frac{|h|^2 t^2}{2!} + \cdots + \frac{|h|^n t^n}{n!} + \cdots \right) = |h| t^2 e^{|h|t}.$$

Now let h tend to zero, and the assertion is obtained.

5.1.5 Remark
Repeated application of this theorem gives the following formulas for the higher derivatives of the function $F(s)$:

$$F^{(n)}(s) = (-1)^n \int_0^\infty t^n f(t)e^{-st}\, dt \text{ for } n \in \{2, 3, \ldots\} \text{ and } s \in \Omega. \tag{4}$$

EXERCISES

501. Find the Laplace transform and its domain for each of the functions:
   a) e^{2t} b) e^{-2t} c) e^{it} d) e^{-it} e) 1
   f) t g) t^2 h) $1 - t^2$ i) $\cos t$ j) $\cos 2t$
   k) $\sin t$ l) $\sin 2t$ m) $\cosh t$ n) $\cosh 2t$ o) $\sinh t$.

502. Find the domain of the Laplace transform for each of the functions
   a) $\dfrac{\sin t}{t}$,   b) $\dfrac{1}{1 + t^2}$,   c) e^{-t^2},   d) e^{t^2}.

503. Suppose that α is a positive number, that β is a nonzero real number, and that n is a positive integer. Use formulas (2) and (4) to compute the integrals

$$\int_0^\infty t^n e^{-\alpha t} \cos \beta t\, dt \quad \text{and} \quad \int_0^\infty t^n e^{-\alpha t} \sin \beta t\, dt.$$

504. In the definition of the Laplace transform, find the function $f_1(t)$ corresponding to the function $f(t) = e^{at}$, $a \in C$.

505. Suppose that a is a complex number and n a positive integer. Show that

$$\mathscr{L}\left[\frac{t^{n-1}e^{at}}{(n-1)!}\right](s) = \frac{1}{(s-a)^n}.$$

Hint.   Use repeated integrations by parts of the last member in (1) with $f(t) = t^{n-1}e^{at}[(n-1)!]^{-1}$.

5.2   INVERSION FORMULA

Theorem 5.2.1 below gives an inversion formula for the Laplace transform. A reader who is not used to considering integrals along curves in the complex plane may follow the proof of Theorem 5.2.1 through formula (b) to see that a function in the class E is uniquely determined by its Laplace transform, read Definition 5.2.2, and then go on to Section 5.3. The integration in formula (1) is along a line in the plane of the variable $s = \sigma + i\omega$ parallel to the imaginary axis, and in the direction of increasing values of ω.

    5.2.1   **Theorem.**   *Suppose that the function $f(t)$ belongs to the class E. Denote its Laplace transform by $F(s)$. Suppose that σ is a real number in the domain of $F(s)$. Then, for all real numbers t,*

$$f(t) = \frac{1}{2\pi i}\int_{\sigma-i\infty}^{\sigma+i\infty} F(s)e^{st}\, ds. \tag{1}$$

    In particular, $f(t)$ is uniquely determined by its Laplace transform $F(s)$.

Proof.   The assumptions imply that, for every number $\sigma + i\omega$ in the domain of $F(s)$,

(a) $$F(\sigma + i\omega) = \int_{-\infty}^{\infty} f(t)e^{-\sigma t}e^{-i\omega t}\, dt.$$

Let σ be a given real number in the domain of $F(s)$. Then $f(t)e^{-\sigma t}$ is a function of t with domain $(-\infty, \infty)$. This function belongs to the class DR. Formula (2) in Corollary 4.4.2 and the equation (a) above show that $f(t)e^{-\sigma t}$ has the Fourier transform $(1/2\pi)F(\sigma + i\omega)$. The inversion formula for Fourier transforms (formula (3) in Corollary 4.4.2) then shows that, for all real numbers t,

$$f(t)e^{-\sigma t} = \frac{1}{2\pi}\int_{-\infty}^{\infty} F(\sigma + i\omega)e^{i\omega t}\, d\omega$$

or

(b) $$f(t) = \frac{1}{2\pi}\int_{-\infty}^{\infty} F(\sigma + i\omega)e^{(\sigma + i\omega)t}\, d\omega.$$

This last equation can be rewritten in the form (1). The claim of uniqueness is an immediate consequence of formula (1). Hence the assertions hold.

5.2.2 Definitions

Formula (1) is called the *inversion formula for Laplace transforms*. The function $f(t)$ is called the *inverse Laplace transform* of the function $F(s)$, and is denoted by $\mathscr{L}^{-1}[F]$:

$$f(t) = \mathscr{L}^{-1}[F(s)](t). \tag{2}$$

In electrical engineering, a rational function is often denoted by $G(s)$ and its inverse Laplace transform by $g(t)$. We shall adhere to this convention.

5.2.3 Theorem.   *Suppose that $G(s)$ is a proper rational function* (i.e. the numerator has lower degree than the denominator) *with complex coefficients. Then $G(s)$ has an inverse Laplace transform $g(t)$. Let t be a given positive number. Denote the poles of the function $G(s)e^{st}$ by $s_1, s_2, \ldots, s_m$. Then $g(t)$ is equal to the sum of the residues at these poles*:

$$g(t) = \sum_{\mu=1}^{m} \operatorname*{Res}_{s=s_\mu} G(s)e^{st} \tag{3}$$

Proof. The function $G(s)$ has a decomposition into partial fractions of the form

(c) $$G(s) = \sum_{\varkappa=1}^{k} \frac{A_\varkappa}{(s - a_\varkappa)^{n_\varkappa}},$$

where the $A_\varkappa$ and the $a_\varkappa$ are complex numbers (each $a_\varkappa$ is one of the numbers s_μ) and the $n_\varkappa$ are positive integers. The operators $\mathscr{L}$ and $\mathscr{L}^{-1}$ are linear (cf. formula (1) in Theorem 5.3.1). It is then sufficient to prove the assertion for one single term in the right member of (c) with $A_\varkappa = 1$. We have

$$\operatorname*{Res}_{s=a} \frac{1}{(s-a)^n} e^{st} = e^{at} \operatorname*{Res}_{s=a} \frac{1}{(s-a)^n} e^{(s-a)t} = e^{at} \operatorname*{Res}_{s=a} \frac{1}{(s-a)^n} \sum_{j=0}^{\varkappa} \frac{(s-a)^j t^j}{j!} = \frac{t^{n-1} e^{at}}{(n-1)!}$$

Theorem 5.2.1 and comparison with the result in Exercise 505 now give the assertion.

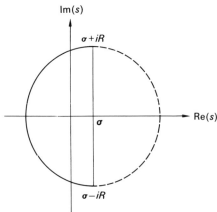

Figure 5.1

Suppose that σ is a real number and R a positive number, such that the poles of $G(s)$ are situated inside the left half-circle of Figure 5.1. It is possible to prove Theorem 5.2.3 by applying the residue theorem to the integral of $G(s)e^{st}$ around the half-circle, letting $R \to \infty$ for fixed σ, applying Lemma 4.4.13, and using formula (1). For $t = 0$ the same procedure shows that $g(0) = \frac{1}{2}g(0^+)$. For $t < 0$ the procedure applied to the right half-circle shows that $g(t) = 0$. These results agree with the fact that $g(t)$ belongs to the class E.

5.2.4 Example

Suppose that a is a nonzero complex number. Find the inverse Laplace transform $g(t)$ of the function $G(s) = s(s^2 + a^2)^{-1}$.

Solution. Decomposition into partial fractions gives

$$\frac{s}{s^2 + a^2} = \frac{1}{2}\left(\frac{1}{s - ia} + \frac{1}{s + ia}\right).$$

Then, for $t > 0$, Theorem 5.2.3 shows that

$$g(t) = \frac{1}{2}\operatorname*{Res}_{s=ia}\frac{e^{st}}{s - ia} + \frac{1}{2}\operatorname*{Res}_{s=-ia}\frac{e^{st}}{s + ia} = \frac{1}{2}e^{iat} + \frac{1}{2}e^{-iat}.$$

Hence, observing that the function $g(t)$ belongs to the class E,

(d)
$$g(t) = \begin{cases} \cos at & \text{for } t > 0 \\ \frac{1}{2} & \text{for } t = 0 \\ 0 & \text{for } t < 0. \end{cases}$$

Formula (d) answers the question in Example 5.2.4 in accordance with Definition 5.2.2. In the following, however, we skip such annotations as in the last two lines in the right member of (d). Then we have the formula (valid also for $a = 0$)

$$\mathscr{L}^{-1}\left[\frac{s}{s^2 + a^2}\right](t) = \cos at \qquad \text{for } a \in C. \tag{4}$$

EXERCISES

*506. Suppose that a is a complex number. Find the inverse Laplace transforms of the following functions by use of Theorem 5.2.3 (a somewhat easier solution is available when Section 5.3 has been studied):

a) $\dfrac{1}{(s - a)^2}$,   b) $\dfrac{a}{s^2 + a^2}$, $a \neq 0$,   c) $\dfrac{1}{s(s + 1)^2}$.

d) Check the answers by computing the Laplace transforms of the functions obtained.

507. Where in the proof of Theorem 5.2.3 is the assumption $t > 0$ applied?

5.3 FURTHER PROPERTIES OF LAPLACE TRANSFORMS. THE CONVOLUTION THEOREM

There are more or less extensive tables of Laplace transforms and inverse Laplace transforms. In computations with Laplace transforms the following theorems (5.3.1–5.3.5) are often useful, even if tables are used. The set of all positive real numbers is denoted by R^+. In Theorem 5.3.1 the reader should visualize the graphs of the functions; e.g. the curve $y = f(t - T)$ is obtained by translation of T units to the right of the curve $y = f(t)$.

5.3.1   Theorem. *Suppose that the functions $f(t)$ and $g(t)$ belong to the class E, that a and b are complex numbers, that α and T are positive numbers, and that n is a positive integer. Then the functions $af(t) + bg(t)$, $e^{at}f(t)$, $f(\alpha t)$, $f(t - T)$ and $t^n f(t)$ belong to the class E. Further, denoting the Laplace transforms of $f(t)$ and $g(t)$ by $F(s)$ and $G(s)$ respectively,*

$$\mathscr{L}[af(t) + bg(t)](s) = aF(s) + bG(s) \qquad a, b \in C \tag{1}$$

$$\mathscr{L}[e^{at}f(t)](s) = F(s - a) \qquad a \in C \tag{2}$$

$$\mathscr{L}[f(\alpha t)](s) = \frac{1}{\alpha} F\left(\frac{s}{\alpha}\right) \qquad \alpha \in R^+ \tag{3}$$

$$\mathscr{L}[f(t - T)](s) = e^{-sT}F(s) \qquad T \in \bar{R}^+ \tag{4}$$

$$\mathscr{L}[t^n f(t)](s) = (-1)^n F^{(n)}(s) \qquad n \in Z^+. \tag{5}$$

Proof. The first assertion and formulas (1) and (2) are readily verified from the definitions of the class E and of the Laplace transform. Formula (3) is obtained by making the substitution $\alpha t = u$ in the integral in

$$\mathscr{L}[f(\alpha t)](s) = \int_0^\infty f(\alpha t)e^{-st}\, dt.$$

Formula (4) is obtained by making the substitution $t - T = u$ in the integral in

$$\mathscr{L}[f(t - T)](s) = \int_T^\infty f(t - T)e^{-st}\, dt.$$

Formula (5) follows from Remark 5.1.5.

5.3.2   Theorem. *Suppose that the function $f(t)$ belongs to the class E, that its derivative $f'(t)$ is equal to a function in the class E except perhaps at finitely many points in each finite interval, that $f(t)$ is continuous for $t > 0$, and that there is a real number γ, such that $f(t)e^{-\gamma t} \to 0$ as $t \to \infty$. Then, denoting the Laplace transform of $f(t)$ by $F(s)$,*

$$\mathscr{L}[f'(t)](s) = sF(s) - f(0^+). \tag{6}$$

Proof. Suppose that s is a number in the domain of $\mathscr{L}[f']$ such that $\text{Re}(s) \geq \gamma$. Then

$$\mathscr{L}[f'(t)](s) = \int_0^\infty f'(t)e^{-st}\, dt = \left[f(t)e^{-st}\right]_{t=0}^{t=\infty} + s\int_0^\infty f(t)e^{-st}\, dt.$$

This proves the assertion.

5.3.3   Corollary. *Suppose that the function $f(t)$ and its successive derivatives $f'(t), f''(t), \ldots, f^{(n-1)}(t)$ satisfy the same conditions as the function $f(t)$ of Theorem 5.3.2, and that its nth derivative $f^{(n)}(t)$ satisfies the same condition*

as the function $f'(t)$ of Theorem 5.3.2. Then, denoting the Laplace transform of
$f(t)$ by $F(s)$,

$$\mathscr{L}[f^{(n)}(t)](s) = s^n F(s) - s^{n-1}f(0^+) - s^{n-2}f'(0^+) - \cdots - f^{(n-1)}(0^+). \quad (7)$$

Proof. Repeated applications of (6) give

$$\mathscr{L}[f''](s) = s\mathscr{L}[f'](s) - f'(0^+)$$
$$\mathscr{L}[f'''](s) = s\mathscr{L}[f''](s) - f''(0^+)$$
$$\cdot$$
$$\cdot$$
$$\cdot$$
$$\mathscr{L}[f^{(n)}](s) = s\mathscr{L}[f^{(n-1)}](s) - f^{(n-1)}(0^+).$$

Elimination of $\mathscr{L}[f']$, $\mathscr{L}[f'']$, ..., $\mathscr{L}[f^{(n-1)}]$ between equation (6) and these
equations gives (7).

5.3.4 Theorem. *Suppose that the function $f(t)$ and its primitive function*
$p(t) = \int_0^t f(u)\,du$ *both belong to the class E, and that there is a real number γ,*
such that $p(t)e^{-\gamma t} \to 0$ as $t \to \infty$. Then, denoting the Laplace transform of
$f(t)$ *by $F(s)$,*

$$\mathscr{L}\left[\int_0^t f(u)\,du\right](s) = \frac{1}{s}F(s). \quad (8)$$

Proof. Application of Theorem 5.3.2 to $p(t)$ gives (8).

5.3.5 Theorem (the *convolution theorem* for Laplace transforms). *Suppose*
that the functions $f(t)$ and $g(t)$ belong to the class E. Then their convolution
$(f * g)(t)$ *belongs to the class E, and the Laplace transforms satisfy the equation:*

$$\mathscr{L}[f * g](s) = \mathscr{L}[f](s)\,\mathscr{L}[g](s). \quad (9)$$

Proof. For $y < 0$ we have $f(y) = 0$ and for $y > t$ we have $g(t - y) = 0$. Hence

$$(f * g)(t) = \begin{cases} \int_0^t f(y)g(t-y)\,dy & \text{for} \quad t > 0 \\ 0 & \text{for} \quad t \leqslant 0. \end{cases}$$

This shows that the convolution is a continuous function, vanishing for $t < 0$.
As in the proof of Theorem 4.5.5 it is seen that the convolution is piecewise continuously differentiable on each finite interval. Suppose that $s = \sigma + i\omega$ is a
complex number that belongs to the domains of both $\mathscr{L}[f]$ and $\mathscr{L}[g]$. Then the
functions $f(t)e^{-st}$ and $g(t)e^{-st}$ both belong to the class DR. Suppose $a > 0$. A
change of the order of integration and the substitution $t - y = u$ give (see Fig. 5.2,
where the integrations are on the shaded domains)

(a) $$\int_0^a (f * g)(t)e^{-st}\,dt = \int_0^a e^{-st}\,dt \int_0^t f(y)g(t-y)\,dy$$

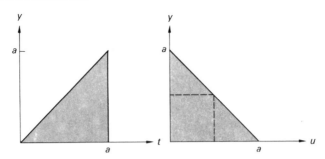

Figure 5.2

$$= \int_0^a f(y)e^{-sy}\,dy \int_y^a g(t-y)e^{-s(t-y)}\,dt$$

$$= \int_0^a f(y)e^{-sy}\,dy \int_0^{a-y} g(u)e^{-su}\,du.$$

It is then seen that

$$\int_0^a |(f * g)(t)e^{-st}|\,dt = \int_0^a |(f * g)(t)|e^{-\sigma t}\,dt$$

$$= \int_0^a e^{-\sigma t}\,dt \left| \int_0^t f(y)g(t-y)\,dy \right|$$

$$\leqslant \int_0^a e^{-\sigma t}\,dt \int_0^t |f(y)|\,|g(t-y)|\,dy$$

$$= \int_0^a |f(y)|e^{-\sigma y}\,dy \int_0^{a-y} |g(u)|e^{-\sigma u}\,du$$

$$\leqslant \int_0^\infty |f(y)|e^{-\sigma y}\,dy \int_0^\infty |g(u)|e^{-\sigma u}\,du.$$

Hence the function $(f*g)(t)e^{-st}$ is integrable on $(0, \infty)$. In summary, the convolution $(f*g)(t)$ has all the properties of a function in the class E. Equation (a) shows that (the integrations in the right member are made in the first quadrant outside the dotted angle in Fig. 5.2)

$$\left| \int_0^a (f * g)(t)e^{-st}\,dt - \mathscr{L}[f](s)\,\mathscr{L}[g](s) \right|$$

$$\leqslant \int_0^{a/2} |f(y)e^{-sy}|\,dy \int_{a/2}^\infty |g(u)e^{-su}|\,du + \int_{a/2}^\infty |f(y)e^{-sy}|\,dy \int_0^\infty |g(u)e^{-su}|\,du.$$

Here the right member tends to zero as $a \to \infty$. Formula (9) follows.

In Theorem 5.3.6 some often used Laplace transforms are collected. It is left to the reader to prove the theorem (Exercise 508). Some of the formulas in Theorem 5.3.6 and part of the discussion in Example 5.3.8 have been given earlier in this text. By means of the theorems in this section (5.3), a large number of Laplace transforms and inverse Laplace transforms can be computed.

5.3.6 Theorem. *Suppose that a is a nonzero complex number and that n is a positive integer. For certain functions f(t) the following list gives the corresponding Laplace transforms:*

$$1 \qquad\qquad \frac{1}{s} \qquad\qquad (10)$$

$$e^{at} \qquad\qquad \frac{1}{s-a} \qquad\qquad (11)$$

$$t^n \qquad\qquad \frac{n!}{s^{n+1}} \qquad\qquad (12)$$

$$\sin at \qquad\qquad \frac{a}{s^2+a^2} \qquad\qquad (13)$$

$$\cos at \qquad\qquad \frac{s}{s^2+a^2} \qquad\qquad (14)$$

$$\sinh at \qquad\qquad \frac{a}{s^2-a^2} \qquad\qquad (15)$$

$$\cosh at \qquad\qquad \frac{s}{s^2-a^2} \qquad\qquad (16)$$

$$\frac{t^{n-1}e^{at}}{(n-1)!} \qquad\qquad \frac{1}{(s-a)^n} \qquad\qquad (17)$$

$$\frac{1}{2a^3}(\sin at - at \cos at) \qquad\qquad \frac{1}{(s^2+a^2)^2} \qquad\qquad (18)$$

$$\frac{t}{2a}\sin at \qquad\qquad \frac{s}{(s^2+a^2)^2} \qquad\qquad (19)$$

5.3.7 Example

Suppose that a is a nonzero complex number. Find the Laplace transforms of the functions $t \sinh at$ and $t \cosh at$.

Solution. Formulas (5), (15) and (16) give

$$\mathscr{L}[t \sinh at](s) = -\frac{d}{ds}\frac{a}{s^2 - a^2} = \frac{2as}{(s^2 - a^2)^2},$$

$$\mathscr{L}[t \cosh at](s) = -\frac{d}{ds}\frac{s}{s^2 - a^2} = \frac{s^2 + a^2}{(s^2 - a^2)^2}.$$

5.3.8 Example

Suppose that $G(s)$ is a proper rational function. Discuss the determination of its inverse Laplace transform $g(t)$.

Discussion. The function $G(s)$ has a decomposition into partial fractions of the form $A(s - a)^{-n}$, where A and a are complex numbers and n a positive integer. We write this observation concisely as

$$G(s) = \sum \frac{A}{(s - a)^n}. \tag{20}$$

Formulas (1) and (17) give

$$g(t) = \sum A \frac{t^{n-1} e^{at}}{(n - 1)!}, \tag{21}$$

where the summation is for the same triples (A, a, n) as in (20). Observe that if the coefficients of $G(s)$ are real and if the nonreal zeros of the denominator are of second order at most, it is convenient to make a decomposition into real partial fractions and to use among other things formulas (2), (13), (14), (18), (19).

Some practical methods are collected in the following auxiliary theorem.

5.3.9 Auxiliary theorem. *Suppose that $G(s)$ is a proper rational function, that $G(s) = P(s)/Q(s)$, where $P(s)$ and $Q(s)$ are two polynomials with real coefficients, and that $g(t)$ is the inverse Laplace transform of $G(s)$. Then:*
1. *To a simple zero α of $Q(s)$ there corresponds in $g(t)$ the term*

$$\frac{P(\alpha)}{Q'(\alpha)} e^{\alpha t}. \tag{22}$$

2. *To two conjugate nonreal simple zeros $\alpha \pm i\omega$ of $Q(s)$ there correspond in $g(t)$ the terms*

$$e^{\alpha t}(A \cos \omega t - B \sin \omega t), \tag{23}$$

where A and B are real numbers determined by

$$A + iB = 2\left[\frac{P(s)}{Q'(s)}\right]_{s = \alpha + i\omega}. \tag{24}$$

3. *If $Q(s) = Q_1(s)Q_2(s)$, where $Q_1(s)$ and $Q_2(s)$ are two polynomials, and if $Q_1(\alpha) = 0$, then*

$$Q'(\alpha) = Q_1'(\alpha)Q_2(\alpha). \tag{25}$$

Proof. **1.** The assertion is an immediate consequence of (21) and of a known expression for certain coefficients in a decomposition into partial fractions.

2. Define A and B by (24). The result in part 1 shows that, to the zeros $\alpha + i\omega$ of $Q(s)$, there correspond in $g(t)$ the terms

$$\frac{P(\alpha + i\omega)}{Q'(\alpha + i\omega)}e^{(\alpha + i\omega)t} + \frac{P(\alpha - i\omega)}{Q'(\alpha - i\omega)}e^{(\alpha - i\omega)t} = 2\,\mathrm{Re}\left[\frac{P(\alpha + i\omega)}{Q'(\alpha + i\omega)}e^{(\alpha + i\omega)t}\right]$$

$$= e^{\alpha t}\,\mathrm{Re}\,[(A + iB)(\cos \omega t + i \sin \omega t)].$$

Hence the assertion holds.

3. This follows immediately from the rule for differentiating a product.

5.3.10 Example

Find the inverse Laplace transform $g(t)$ of the function

$$G(s) = \frac{s}{(s + 3)(s^2 + 4s + 5)}.$$

Solution. We apply the auxiliary theorem. To the zero -3 of the denominator there corresponds in $g(t)$ the term

$$\left[\frac{s}{s^2 + 4s + 5}e^{st}\right]_{s=-3} = \frac{-3}{9 - 12 + 5}e^{-3t} = -\frac{3}{2}e^{-3t}.$$

To the zeros $-2 \pm i$ there correspond in $g(t)$ the terms

$$e^{-2t}(A \cos t - B \sin t)$$

where

$$A + iB = 2\left[\frac{s}{(s + 3)(2s + 4)}\right]_{s=-2+i} = 2\frac{-2 + i}{(1 + i)2i}\frac{1 - i}{1 - i} = \frac{-1 + 3i}{2i} = \frac{3 + i}{2};$$

$$A + \tfrac{3}{2}, \qquad B = \tfrac{1}{2}.$$

Hence

$$g(t) = -\tfrac{3}{2}e^{-3t} + e^{-2t}(\tfrac{3}{2}\cos t - \tfrac{1}{2}\sin t).$$

EXERCISES

508. Prove Theorem 5.3.6.

509. Find the Laplace transforms of the functions
  a) $t^4 + 2$ b) $te^t - 1$
  c) $\sin 3t - \cos 5t$ d) $(1 - t^2)e^{2t}$.

510. Suppose that α and ω are nonzero real numbers. Find the Laplace transforms of the functions

   a) $e^{-\alpha t} \sin \omega t$, b) $e^{-\alpha t} \cos \omega t$.

511. Consider the function $u(t)$ defined by: $u(t) = 0$ for $t < 0$, $u(0) = \frac{1}{2}$, $u(t) = 1$ for $t > 0$. This function $u(t)$ is called the *unit step function*. Suppose that T is a positive number. The function $u(t - T)$ is called a *delayed unit step function*. The function $u(t) - u(t - T)$ is called a *rectangular pulse function*. Sketch the graphs of the three functions $u(t)$, $u(t - T)$, $u(t) - u(t - T)$. Find the Laplace transforms of the three functions.

512. Suppose that T is a positive number. Suppose that $f_T(t)$ and $f(t)$ are two functions in the class E with the properties: $f_T(t) = 0$ for $t > T$, $f(t) = f_T(t)$ for $0 < t < T$, $f(t + T) = f(t)$ for $t > 0$. Denote the Laplace transforms of the functions by $F_T(s)$ and $F(s)$ respectively. Prove the formula

$$F(s) = \frac{F_T(s)}{1 - e^{-Ts}}.$$

513. Use the result in Exercise 512 to compute the Laplace transforms of the functions
   a) $f(t) = |\sin t|$ (a full-wave rectified sine oscillation).
   b) $f(t) = \max (\sin t, 0)$ (a half-wave rectified sine oscillation).

514. Find the Laplace transform of the function $f(t) = 1/\sqrt{t}$.
   Hint. Use formula (9) of Example 4.4.12.

515. Suppose that a and b are complex numbers. Find the inverse Laplace transforms of the functions (the auxiliary theorem 5.3.9 may be applied):

   a) $\dfrac{s + 4}{s(s^2 + 4s + 13)}$, b) $\dfrac{s^2}{(s^2 + 1)(s^2 + 2s + 10)}$,

   c) $\dfrac{s - a}{(s - a)^4 - 1}$, d) $\dfrac{2s^2 + 1}{s^3 + s}$, e) $\dfrac{d^n}{ds^n} \dfrac{a + bs}{s^2 + 1}$.

*516. Suppose that $g(t)$ is the inverse Laplace transform of a proper rational function $G(s)$, whose poles are all situated in the half-plane $\text{Re}(s) < 0$. Find $\lim_{t \to \infty} g(t)$.

517. Suppose that the function $f(t)$ belongs to the class E, and denote its Laplace transform by $F(s)$. Consider only real values of s.
   a) Show that the limit $\lim_{s \to \infty} sF(s)$ exists and is equal to $f(0^+)$.
   b) Show that if the limit $\lim_{t \to \infty} f(t)$ exists, then the limit $\lim_{s \to 0^+} sF(s)$ also exists, and these two limits are equal.
   c) Illustrate properties a) and b) by verifying them for the functions $f(t) = 1$, $f(t) = e^{-t}$, $f(t) = \sin t$, $f(t) = \cos t$.
   Hint. a) Show that both integrals in the equation

$$sF(s) - f(0^+) = \int_0^{1/\sqrt{s}} [f(t) - f(0^+)] s e^{-st} dt$$

$$+ \int_{1/\sqrt{s}}^{\infty} [f(t) - f(0^+)] s e^{-st} dt$$

tend to zero, as $s \to \infty$. b) Set $f(\infty) = \lim_{t \to \infty} f(t)$. Write $sF(s) - f(\infty)$ as an integral, and continue as in a).

518. Does the statement of Theorem 5.3.2 contain any superfluous assumption?

519. Suppose that x is a positive number. Show that

$$\mathscr{L}\left[\frac{x}{2\sqrt{\pi t^3}} e^{-x^2/(4t)}\right](s) = e^{-x\sqrt{s}}.$$

(This result will be used in Problem 8.4.5.)

Hint. Suppose $k > 0$. Use the substitution $\sqrt{y} - k/\sqrt{y} = z$ and formula (9) of Example 4.4.12 to evaluate the integral

$$I = \int_0^\infty y^{-3/2} e^{-y-k^2/y} \, dy.$$

It will be found that

$$I = \frac{1}{k} e^{-2k} \int_{-\infty}^\infty e^{-z^2} \, dz = \frac{\sqrt{\pi}}{k} e^{-2k}.$$

Then set $y = st$ and $k^2 = x^2 s/4$, and the assertion follows.

5.4 APPLICATIONS TO ORDINARY DIFFERENTIAL EQUATIONS

The term "ordinary" in the section heading means that there is just one independent variable. In Chapters 7 and 8 we shall study "partial" differential equations, which have two or more independent variables.

5.4.1 Introductory material

In this section (5.4) we shall discuss application of the Laplace transform to determine the solutions of an *nth-order linear differential equation with constant coefficients.* Such a differential equation is of the form

$$y^{(n)} + a_{n-1}y^{(n-1)} + \cdots + a_1 y' + a_0 y = f(t), \qquad t \in (a, b), \tag{1}$$

where n is a positive integer, $a_{n-1}, \ldots, a_0$ are real numbers, (a, b) is a finite or infinite open interval, and $f(t)$ is a continuous real-valued function with domain (a, b). We say that the differential equation is *given* on the interval (a, b). When no interval is given in the context of a differential equation, it is understood that the differential equation is given in the largest open interval where $f(t)$ is defined. A *solution* or a *particular solution* of (1) is a complex-valued function $y(t)$ that is differentiable at least n times and that together with its derivatives $y'(t), y''(t), \ldots,$ $y^{(n)}(t)$ satisfies the differential equation in (1) on the interval (a, b). The *general solution* of (1) is the set of all its solutions. To *solve* (1) is to determine its general solution. If $f(t) = 0$ for $a < t < b$, the differential equation is *homogeneous*, otherwise it is *nonhomogeneous*. If (1) is nonhomogeneous, the differential equation

$$y^{(n)} + a_{n-1}y^{(n-1)} + \cdots + a_1 y' + a_0 y = 0 \tag{2}$$

is called its *associated homogeneous equation*. The algebraic equation

$$r^n + a_{n-1}r^{n-1} + \cdots + a_1 r + a_0 = 0 \tag{3}$$

in the unknown r is called its *auxiliary equation*.

In several places we shall rely on the following lemma on existence and uniqueness of the solution of a certain initial-value problem. For a proof of the lemma, see [18], Theorem 8.1. For future use we allow the coefficients in (4) to be variable.

In applications in physics some quantity y is supposed to vary with time t for $t \geq t_0$ according to a differential equation (4), the values (5) being given; the function values $y(t)$ for $t > t_0$ are asked for. This accounts for the term "initial-value problem".

5.4.2   Lemma. *Suppose that (a, b) is a finite or infinite open interval, that $a_{n-1}(t), \ldots, a_1(t), a_0(t)$ and $f(t)$ are continuous real-valued functions on (a, b), that $a < t_0 < b$, and that $y_0, y_1, \ldots, y_{n-1}$ are complex numbers. Then the differential equation*

$$y^{(n)} + a_{n-1}(t)y^{(n-1)} + \cdots + a_1(t)y' + a_0(t)y = f(t), \qquad t \in (a, b), \tag{4}$$

has one and only one solution $y(t)$ on (a, b) such that

$$y(t_0) = y_0, \; y'(t_0) = y_1, \ldots, y^{(n-1)}(t_0) = y_{n-1}. \tag{5}$$

5.4.3 Problem

Suppose that a_1 and a_0 are real numbers. Solve the differential equation

$$y'' + a_1 y' + a_0 y = 0. \tag{6}$$

Solution. **1.** Suppose that $y(t)$ is a solution of (6). Set $y_0 = y(0)$ and $y_1 = y'(0)$. Suppose that Theorems 5.3.1–5.3.3 are applicable. They then give, setting $Y(s) = \mathscr{L}[y](s)$,

$$s^2 Y(s) - s y_0 - y_1 + a_1 s Y(s) - a_1 y_0 + a_0 Y(s) = 0,$$

(a)
$$Y(s) = \frac{s y_0 + y_1 + a_1 y_0}{s^2 + a_1 s + a_0}.$$

First assume that the auxiliary equation $r^2 + a_1 r + a_0 = 0$ has different roots r_1 and r_2. Decomposition of the right member of (a) into partial fractions gives two constants c_1 and c_2 such that

(b)
$$Y(s) = \frac{c_1}{s - r_1} + \frac{c_2}{s - r_2}.$$

Then, by Theorems 5.2.1, 5.3.1 and 5.3.6, we have at least for $t > 0$

$$y(t) = c_1 e^{r_1 t} + c_2 e^{r_2 t}. \tag{7}$$

Now assume that the auxiliary equation has a double root r_1. Then the formulas (b) and (7) are replaced by

$$Y(s) = \frac{c_1}{(s - r_1)^2} + \frac{c_2}{s - r_1},$$

$$y(t) = (c_1 t + c_2)e^{r_1 t}. \tag{8}$$

2. Let c_1 and c_2 be given complex numbers. It is readily verified that the functions (7) and (8) respectively are solutions of the differential equation (6) on $(-\infty, \infty)$ in the two cases considered in part 1.

3. In the deduction of formulas (7) and (8) we introduced an extra assumption (viz. that Theorems 5.3.1–5.3.3 are applicable). Then (6) might have solutions not of any of the forms (7) and (8). However, Lemma 5.4.2 shows that the differential equation (6) has one and only one solution $y(t)$ with prescribed values of $y(0)$ and $y'(0)$. The coefficients c_1 and c_2 in (7) and (8) respectively can be so chosen that $y(0)$ and $y'(0)$ equal these prescribed values. Then, interpreting c_1 and c_2 in (7) and (8) as complex parameters, it is seen that (6) has the general solution (7) if the auxiliary equation has different roots r_1 and r_2, and the general solution (8) if the auxiliary equation has a double root r_1.

5.4.4 A generalization
Consider the differential equation (2). Suppose that its auxiliary equation (3) has the roots $r_1, r_2, \ldots, r_k$, with the multiplicities $m_1, m_2, \ldots, m_k$ respectively. (Then $m_1 + \cdots + m_k = n$.) Just as in Problem 5.4.3 it is seen that (2) has the general solution

$$y(t) = P_1(t)e^{r_1 t} + P_2(t)e^{r_2 t} + \cdots + P_k(t)e^{r_k t}, \tag{9}$$

where each $P_x(t)$ is a polynomial of degree $m_x - 1$ with coefficients interpreted as complex parameters.

5.4.5 Problem
Discuss solution of the differential equation (1).

Discussion. Suppose that $a < 0 < b$. (This can always be achieved by a linear change of the independent variable.) Restrict attention to the interval $(0, b)$. Let $y_p(t)$ be the particular solution of (1) for which $y_p(0) = y_p'(0) = \cdots = y_p^{(n-1)}(0) = 0$. (The function $y_p(t)$ exists and is unique by Lemma 5.4.2.) Replace y by y_p in (1), and suppose that Theorems 5.3.1–5.3.3 are applicable. We then get, setting $Y_p(s) = \mathscr{L}[y_p](s)$ and $F(s) = \mathscr{L}[f](s)$,

$$(s_n + a_{n-1}s^{n-1} + \cdots + a_1 s + a_0)Y_p(s) = F(s).$$

Let

$$G(s) = \frac{1}{s^n + a_{n-1}s^{n-1} + \cdots + a_1 s + a_0}$$

and

$$g(t) = \mathscr{L}^{-1}[G(s)](t). \tag{10}$$

Then

$$Y_p(s) = F(s)G(s).$$

Hence, by the convolution theorem for Laplace transforms (Theorem 5.3.5),

$$y_p(t) = (f * g)(t) = \int_0^t f(\tau)g(t - \tau) \, d\tau. \tag{11}$$

It can be proved (see Exercises 522–523), using the notations in (9), that (1) has the general solution for $0 < t < b$:

$$y(t) = P_1(t)e^{r_1 t} + \cdots + P_k(t)e^{r_k t} + \int_0^t f(\tau)g(t - \tau) \, d\tau. \tag{12}$$

In practical applications one gets an analytic expression for the integral in (12) such that the general solution on all of (a, b) is obtained.

5.4.6 Example
Solve the differential equation

$$y'' + y = 2 \cos^{-3} t, \qquad t \in \left(-\frac{\pi}{2}, \frac{\pi}{2}\right).$$

Solution. The auxiliary equation is $r^2 + 1 = 0$ with the roots $+i$ and $-i$. The expression (9) then is $y(t) = C_1 e^{it} + C_2 e^{-it}$ or, in trigonometric form,

$$y(t) = c_1 \cos t + c_2 \sin t.$$

The function (10) is

$$g(t) = \mathcal{L}^{-1}\left[\frac{1}{s^2 + 1}\right](t) = \sin t.$$

Suppose $0 < t < \pi/2$. The function (11) is

$$y_p(t) = \int_0^t \frac{2}{\cos^3 \tau} \sin(t - \tau) \, d\tau$$

$$= 2 \sin t \int_0^t \frac{d\tau}{\cos^2 \tau} - 2 \cos t \int_0^t \frac{\sin \tau}{\cos^3 \tau} \, d\tau$$

$$= 2 \sin t \tan t - \cos t \left[\frac{1}{\cos^2 \tau}\right]_{\tau = 0}^{\tau = t}$$

$$= \frac{2 \sin^2 t}{\cos t} - \frac{1}{\cos t} + \cos t = \frac{1}{\cos t} - \cos t.$$

This function is also a solution for $-\pi/2 < t \leqslant 0$. We then have the general solution (we let $-\cos t$ be absorbed into the term $c_1 \cos t$):

$$y(t) = c_1 \cos t + c_2 \sin t + \frac{1}{\cos t}.$$

The next example demonstrates that the Laplace transform technique can be used to solve certain initial-value problems for differential equations without first determining the general solution. In the following solution only the computations are shown; it is left to the reader to supply details.

5.4.7 Example

Solve the initial-value problem

$$y'' - y = 2e^t, \qquad y(0) = 2, \qquad y'(0) = 1.$$

Solution. We have, setting $Y(s) = \mathcal{L}[y](s)$,

$$s^2 Y(s) - 2s - 1 - Y(s) = \frac{2}{s-1},$$

$$(s^2 - 1) Y(s) = \frac{2}{s-1} + 2s + 1 = \frac{2s^2 - s + 1}{s-1},$$

$$Y(s) = \frac{2s^2 - s + 1}{(s+1)(s-1)^2} = \frac{1}{s+1} + \frac{1}{s-1} + \frac{1}{(s-1)^2},$$

and finally

$$y(t) = e^{-t} + e^t + te^t.$$

EXERCISES

520. Solve the differential equations
   a) $2y'' - 4y' + y = 0$
   b) $y''' - 5y'' + 8y' - 4y = 0$
   c) $y'' - 10y' + 61y = 0$
   d) $y^{(4)} - 2y''' + 2y'' = 0$
   e) $y'' + y = t \cos t$
   f) $y'' - 2y' - 3y = \sin 2t + 4 \cos 2t$
   g) $y''' + 4y'' + 5y' = t(1 + e^{-t})$
   h) $y'' + 2y' + y = e^{-t}/t^2, \qquad t > 0$
   i) $y'' + 4y = 1/\cos 2t, \qquad |t| < \pi/4$
   j) $y'' + y = \cot^2 t, \qquad 0 < t < \pi$
   k) $y'' - 5y' + 6y = t \log t, \qquad t > 0.$

521. Solve the initial-value problems
   a) $y'' + 2y' + 2y = 0, \qquad y(0) = 2, \qquad y'(0) = -1$
   b) $y'' + y = 3 \sin 2t, \qquad y(0) = 1, \qquad y'(0) = -2$
   c) $y'' - 2y' + y = 0, \qquad y(0) = y_0, \qquad y'(0) = y_1.$

*522. Suppose that $y_p(t)$ is a solution of (1). Show that if $y_p(t)$ is added to each solution of (2), then the general solution of (1) is obtained.

523. Show that (11) is a solution of (1) for $0 < t < b$.

Hint. First show, for $0 < t < b$, that $g(t)$ is the solution of (2) that satisfies the conditions $y(0) = y'(0) = \cdots = y^{(n-2)}(0) = 0$, $y^{(n-1)}(0) = 1$. Then verify by differentiations and substitution that $y_p(t) = \int_0^t f(\tau)g(t - \tau)d\tau$ satisfies (1) for $0 < t < b$.

CHAPTER 6

BESSEL FUNCTIONS

6.1 THE GAMMA FUNCTION

6.1.1 Definition

Suppose that x is a positive number. Then $t^{x-1}e^{-t}$ is a function of t, integrable on the infinite interval $(0, \infty)$. Definition of the *gamma function* for $x > 0$ and notation for this function are given by the equation

$$\Gamma(x) = \int_0^\infty t^{x-1}e^{-t}\, dt \qquad \text{for} \quad x > 0. \qquad (1)$$

6.1.2 Properties

It is seen that

$$\Gamma(1) = 1. \qquad (2)$$

Further, the partial integration

$$\Gamma(x + 1) = \int_0^\infty t^x e^{-t}\, dt = -\left[t^x e^{-t}\right]_{t=0}^{t=\infty} + x \int_0^\infty t^{x-1}e^{-t}\, dt$$

gives for $x > 0$

$$\Gamma(x + 1) = x\Gamma(x).$$

Repeated application of this equation shows that, for $x > 0$ and each positive integer n,

$$\Gamma(x + n) = (x + n - 1)(x + n - 2)\cdots(x + 1)x\Gamma(x).$$

This equation and equation (2) show that

$$\Gamma(n + 1) = n! \qquad \text{for} \quad n \in Z^+. \qquad (3)$$

6.1.3 Definition

We shall also give a definition of $\Gamma(x)$ for negative nonintegral values of x. To this end observe that if $n \in Z^+$, $x < 0$ and $x + n > 0$, then everything in the equation for $\Gamma(x + n)$ above except $\Gamma(x)$ is defined. The definition of the *gamma function* is then completed by the equation

$$\Gamma(x) = \frac{\Gamma(x + n)}{x(x + 1)\cdots(x + n - 1)} \qquad \text{for} \quad x < 0, \quad n \in Z^+, \quad x + n > 0. \qquad (4)$$

6.1.4 Properties

It is left to the reader (Exercise 601) to show that this definition is consistent (i.e. that different values of n give the same value for $\Gamma(x)$), that the following recursion formula holds for every number x in the domain of the gamma function:

$$\Gamma(x + 1) = x\Gamma(x) \qquad \text{for} \quad x \in D_\Gamma, \tag{5}$$

and that $|\Gamma(x)| \to \infty$ as x tends to zero or to a negative integer. Figure 6.1 shows the graph of the gamma function.

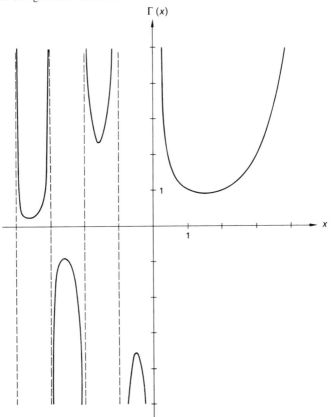

Figure 6.1

The substitution $t = y^2$ gives

$$\Gamma(\tfrac{1}{2}) = \int_0^\infty t^{-\frac{1}{2}} e^{-t} \, dt = \int_0^\infty \frac{1}{y} e^{-y^2} 2y \, dy.$$

Hence, by formula (9) of Example 4.4.12

$$\Gamma(\tfrac{1}{2}) = \sqrt{\pi}. \tag{6}$$

Certain books with mathematical tables give information on the values of the gamma function for $1 < x < 2$. For other values of x, formula (5) can be used.

EXERCISES

601. Prove the three properties stated in the comments concerning formula (4).

602. Suppose that n is a positive integer. Find

     a) $\Gamma(\tfrac{1}{2} + n)$,   b) $\Gamma(\tfrac{1}{2} - n)$,   c) $\lim\limits_{n \to \infty} \Gamma(\tfrac{1}{2} - n)$.

603. Suppose that $x > 0$ and $a > 0$. Express the following integrals in terms of the gamma function:

     a) $\displaystyle\int_0^\infty t^{x-1} e^{-at}\, dt$,   b) $\displaystyle\int_0^\infty e^{-t^a}\, dt$,   c) $\displaystyle\int_0^1 \left(\log \frac{1}{t}\right)^{a-1} dt$.

*604. Show that, for $x > 0$,

$$\Gamma'(x) = \int_0^\infty t^{x-1} e^{-t} \log t\, dt.$$

6.2 THE BESSEL DIFFERENTIAL EQUATION. BESSEL FUNCTIONS
6.2.1 Definition
Suppose that p is a real number. Then the differential equation

$$x^2 y'' + xy' + (x^2 - p^2)y = 0, \qquad x > 0, \quad p \in R, \tag{1}$$

is called the *Bessel differential equation of order p*.

To every real number p there corresponds a differential equation (1). To the numbers p and $-p$ there corresponds the same differential equation. If $y_1(x)$ and $y_2(x)$ are two solutions of (1), linearly independent on the infinite interval $(0, \infty)$, then the general solution on $(0, \infty)$ is given by

$$y = c_1 y_1(x) + c_2 y_2(x) \tag{2}$$

(see Exercise 608; here and in the following, c_1 and c_2 denote complex parameters.) The restriction $x > 0$ in (1) is essential: We shall see in Section 6.8 that (2) is not the general solution on any larger interval (the zero of the coefficient of y'' forms a "barrier"). A simpler example where the same phenomenon occurs is the differential equation

$$x^2 y'' + xy' - y = 0, \qquad x > 0.$$

Its general solution

$$y = c_1 x + c_2/x$$

is not defined for $x = 0$.

An argument that includes a fairly long computation, introduces the function (a) below, defined by a series, as a probable solution of (1). We omit this argument here (the interested reader can consult Exercise 609). We only give a lemma to the effect that (a) is a solution. For the restriction in the lemma that p is not a negative integer, we have so far two reasons: The denominator in (a) is not defined for $-p \in Z^+$ and sufficiently small values of k; the analysis in Exercise 609 makes it probable that the case $-p \in Z^+$ is a special one.

6.2.2 Lemma. *Suppose that p is a real number different from* $-1, -2, -3, \ldots$. *Then the Bessel differential equation (1) is satisfied by the function*

(a) $$y = \sum_{k=0}^{\infty} \frac{(-1)^k}{k! \Gamma(k + p + 1)} \left(\frac{x}{2}\right)^{2k+p}, \qquad x > 0.$$

Proof. Because $\Gamma(k + p + 1) \to \infty$ as $k \to \infty$, the series converges for each $x > 0$. Hence the function (a) has the domain $(0, \infty)$. Set

$$a_k = \frac{(-1)^k}{k! \Gamma(k + p + 1) 2^{2k+p}} \qquad \text{for} \quad k \in N.$$

Differentiations term by term are legitimate:

$$y = \sum_{k=0}^{\infty} a_k x^{2k+p},$$

$$y' = \sum_{k=0}^{\infty} (2k + p) a_k x^{2k+p-1},$$

$$y'' = \sum_{k=0}^{\infty} (2k + p)(2k + p - 1) a_k x^{2k+p-2}.$$

Substitution into the left member of (1) gives as the coefficient for x^p

$$p(p - 1)a_0 + pa_0 - p^2 a_0 = 0,$$

and as the coefficient for x^{2k+p}, $k > 0$,

$$(2k + p)(2k + p - 1)a_k + (2k + p)a_k + a_{k-1} - p^2 a_k$$
$$= (2k + p)^2 a_k + a_{k-1} - p^2 a_k = 4k(k + p)a_k + a_{k-1} = 0.$$

Hence the assertion holds.

6.2.3 Definition

Suppose again that p is a real number but not a negative integer. The notation for and the definition of the *Bessel function of order p* are given by the equation

$$J_p(x) = \sum_{k=0}^{\infty} \frac{(-1)^k}{k! \Gamma(k + p + 1)} \left(\frac{x}{2}\right)^{2k+p}, \qquad x > 0, \quad p \in R, \quad -p \notin Z^+. \qquad (3)$$

(In the literature the function (3) is also called the Bessel function of first kind of order p and the cylindrical function of order p.)

6.2.4 Theorem. *Suppose that p is a nonintegral real number. Then the Bessel differential equation of order p has the general solution*

$$y = c_1 J_p(x) + c_2 J_{-p}(x), \qquad x > 0, \quad p \in R, \quad p \notin Z. \tag{4}$$

Proof. Suppose $p > 0$. Lemma 6.2.2 shows that $J_p(x)$ and $J_{-p}(x)$ are solutions. Further they are linearly independent, for $J_p(x) \to 0$ and $|J_{-p}(x)| \to \infty$ as $x \to 0$. Hence (4) is the general solution (cf. Exercise 608). In the case $p < 0$, the assertion is proved analogously.

EXERCISES

605. Solve the differential equation

$$x^2 y'' + x y' + (x^2 - 9/4)y = 0, \qquad x > 0.$$

606. Show that the differential equation $y'' + xy = 0$, $x > 0$, has the general solution

$$y = c_1 x^{\frac{1}{2}} J_{1/3}(\tfrac{2}{3} x^{3/2}) + c_2 x^{\frac{1}{2}} J_{-1/3}(\tfrac{2}{3} x^{3/2}).$$

607. Show that the differential equation $xy'' + (x^2 - 2)y = 0$, $x > 0$, has the general solution

$$y = c_1 x^{\frac{1}{2}} J_{3/2}(x) + c_2 x^{\frac{1}{2}} J_{-3/2}(x).$$

*608. Use Lemma 5.4.2 to show that, under the hypotheses in the text, (2) is the general solution of (1).

609. The substitution $y = x^p z$ transforms the Bessel differential equation (1) into a new differential equation, that is satisfied by a power series of the form $z = \sum_{k=0}^{\infty} c_k x^{2k}$. This power series is uniquely determined up to a constant factor. Use this to determine a solution of the Bessel differential equation.
    Hint. The substitution gives $xz'' + (2p + 1)z' + xz = 0$. Substitution of the power series and reduction yields

$$\sum_{k=1}^{\infty} [4k(p + k)c_k + c_{k-1}]x^{2k-1} = 0.$$

First suppose that p is not a negative integer. Then we get for each $k \in Z^+$

$$c_k = -\frac{c_{k-1}}{2^2 k(p + k)}, \qquad c_k = \frac{(-1)^k c_0}{2^{2k} k!(p + k)(p + k - 1) \cdots (p + 1)}.$$

Set $c_0 = 2^{-p}[\Gamma(p + 1)]^{-1}$ (this is an established convention). Then (a) is obtained. Now suppose that p is a negative integer. Set $n = -p$. Then

$$\sum_{k=1}^{\infty} [4k(k - n)c_k + c_{k-1}]x^{2k-1} = 0,$$

$c_{n-1} = c_{n-2} = \cdots = c_0 = 0$, and for each $k \in Z^+$

$$c_{n+k} = -\frac{c_{n+k-1}}{2^2(n+k)k}, \qquad c_{n+k} = \frac{(-1)^k c_n}{2^{2k}(n+k)\cdots(n+1)k(k-1)\cdots 1}.$$

Set $c_n = 2^{-n}(n!)^{-1}$ (also an established convention). Then

$$c_{n+k} = \frac{(-1)^k}{2^{2k+n}(n+k)!k!}, \qquad y = \sum_{k=0}^{\infty} \frac{(-1)^k}{k!(k+n)!} \left(\frac{x}{2}\right)^{2k+n}$$

This is, however, the series that is obtained by setting $p = n$ in (a). Hence, for negative integral values of p, the procedure does not lead to a new solution of (1).

6.3 SOME PARTICULAR BESSEL FUNCTIONS

6.3.1 Bessel functions of integral order

Suppose that n is a nonnegative integer, and consider the Bessel differential equation of order n:

$$x^2 y'' + xy' + (x^2 - n^2)y = 0, \qquad x > 0, \quad n \in N. \tag{1}$$

Lemma 6.2.2 shows that the Bessel function

$$J_n(x) = \sum_{k=0}^{\infty} \frac{(-1)^k}{k!(k+n)!} \left(\frac{x}{2}\right)^{2k+n}, \qquad x > 0, \quad n \in N, \tag{2}$$

is a solution of (1). In particular,

$$J_0(x) = \sum_{k=0}^{\infty} \frac{(-1)^k}{(k!)^2} \left(\frac{x}{2}\right)^{2k}$$

$$= 1 - \frac{x^2}{2^2} + \frac{x^4}{2^2 \cdot 4^2} - \frac{x^6}{2^2 \cdot 4^2 \cdot 6^2} + \cdots, \qquad x > 0. \tag{3}$$

Figure 6.2 shows the graphs of some Bessel functions of integral order. Some collections of tables give information on functional values of some Bessel functions.

It is seen that the series in (2) is convergent for all real values of x, that its sum is an even function or an odd function according as n is even or odd, and that the sum satisfies the differential equations in (1) for all real values of x. However, in the three formulas above we have written $x > 0$ instead of $x \in R$, this with rexpect to the general solution of (1) that will be discussed in Section 6.8. (We call attention to the analysis in Exercise 609 that indicates that the counterpart of Theorem 6.2.4 for integral orders may be complicated.) We shall show that the Bessel functions of orders $\frac{1}{2}$ and $-\frac{1}{2}$ are elementary functions.

6.3.2 Bessel Functions of orders $\frac{1}{2}$ and $-\frac{1}{2}$

The result in Exercise 602a shows that

$$J_{\frac{1}{2}}(x) = \sum_{n=0}^{\infty} \frac{(-1)^n}{n!\Gamma(n+\frac{3}{2})} \left(\frac{x}{2}\right)^{2n+\frac{1}{2}} = \sqrt{\frac{2}{\pi x}} \sum_{n=0}^{\infty} \frac{(-1)^n 2^{n+1}}{n! \cdot 1 \cdot 3 \cdots (2n+1)} \left(\frac{x}{2}\right)^{2n+1}$$

$$= \sqrt{\frac{2}{\pi x}} \sum_{n=0}^{\infty} \frac{(-1)^n x^{2n+1}}{(2n+1)!}.$$

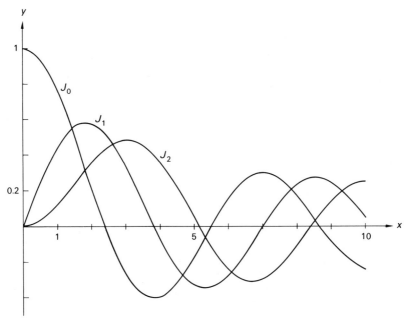

Figure 6.2

Hence

$$J_{\frac{1}{2}}(x) = \sqrt{\frac{2}{\pi x}}\, \sin x. \tag{4}$$

Analogously

$$J_{-\frac{1}{2}}(x) = \sum_{n=0}^{\infty} \frac{(-1)^n}{n!\Gamma(n+\frac{1}{2})} \left(\frac{x}{2}\right)^{2n-\frac{1}{2}} = \sqrt{\frac{2}{\pi x}} \sum_{n=0}^{\infty} \frac{(-1)^n 2^n}{n! \cdot 1 \cdot 3 \cdots (2n-1)} \left(\frac{x}{2}\right)^{2n}$$

$$= \sqrt{\frac{2}{\pi x}} \sum_{n=0}^{\infty} \frac{(-1)^n x^{2n}}{(2n)!}.$$

Hence

$$J_{-\frac{1}{2}}(x) = \sqrt{\frac{2}{\pi x}}\, \cos x. \tag{5}$$

EXERCISES

610. Show that the function $xJ_1(x)$ is a solution of the differential equation

$$xy'' - y' + xy = 0, \quad x > 0.$$

611. Suppose that $m > 0$. Find a solution of the differential equation

$$x^2 y'' + 3xy' + (m^2 x^2 + 1)y = 0, \quad x > 0.$$

Hint. Make the substitution $y = z/x$.

612. Find a solution of the differential equation $xy'' + 5y' + xy = 0, \quad x > 0$.
Hint. Make the substitution $y = x^m z$ with m conveniently chosen.

613. Suppose that m is a nonzero real number. Find a solution of the differential equation

$$xy'' + y' + m^2 x^{2m-1} y = 0, \quad x > 0.$$

Hint. Make the substitution $x^m = t$.

614. Suppose $x > 0$. Determine the derivative $J'_{1/2}(x)$ in terms of the functions $J_{1/2}(x)$ and $J_{-1/2}(x)$.

*615. Suppose that $n \in N$ and $x > 0$. Show that

$$|J_n(x)| \leqslant 2^{-n}(n!)^{-1} x^n e^{x^2/4}.$$

(Better estimates for large values of x are given in Sections 6.5 and 6.7.)

616. Find the Laplace transform of the Bessel function $J_0(x)$.
Hint. Use the fact that $J_0(x)$ satisfies a certain differential equation with certain initial values $J_0(0)$ and $J'_0(0)$. It follows that its Laplace transform satisfies a certain first-order ordinary differential equation. To choose the correct particular solution from the general solution of this equation, apply Exercise 517a.

617. Show that for $x > 0$

a) $J_0(x) = \dfrac{1}{2\pi} \displaystyle\int_0^{2\pi} e^{ix\cos\theta}\, d\theta = \dfrac{1}{\pi} \displaystyle\int_0^{\pi} \cos(x\cos\theta)\, d\theta,$

b) $\dfrac{\sin x}{x} = \displaystyle\int_0^{\pi/2} J_0(x\cos\theta)\cos\theta\, d\theta.$

Hint. Integrate the power-series expansions of the integrands term by term.

6.4 RECURSION FORMULAS FOR THE BESSEL FUNCTIONS

We shall state and prove some recursion formulas for the Bessel functions. In order to avoid some unnecessary restrictions, we also introduce Bessel functions for negative integral values of p. We do this by the following definition (which is natural because of a result in Exercise 609 and the fact that the recursion formulas become valid for all real values of p).

6.4.1 Definition

Suppose that n is a positive integer. The *Bessel function of order* $-n$ is defined by

$$J_{-n}(x) = (-1)^n J_n(x), \quad x > 0, \quad n \in Z^+. \tag{1}$$

6.4.2 Theorem. *Suppose that p is a real number. Then the following recursion formulas hold for the Bessel functions of orders p, $p - 1$, $p + 1$:*

$$\frac{2p}{x} J_p(x) = J_{p-1}(x) + J_{p+1}(x), \quad x > 0, \quad p \in R, \tag{2}$$

$$2J'_p(x) = J_{p-1}(x) - J_{p+1}(x), \qquad x > 0, \quad p \in R. \tag{3}$$

Proof. We have for $-p \notin N$

$$\frac{d}{dx}[x^p J_p(x)] = \frac{d}{dx} \sum_{k=0}^{\infty} \frac{(-1)^k}{k! \Gamma(k+p+1)} \frac{x^{2k+2p}}{2^{2k+p}}$$

$$= \sum_{k=0}^{\infty} \frac{(-1)^k 2(k+p)}{k!(k+p)\Gamma(k+p)} \frac{x^{2k+2p-1}}{2^{2k+p}}$$

$$= x^p \sum_{k=0}^{\infty} \frac{(-1)^k}{k! \Gamma(k+p)} \frac{x^{2k+p-1}}{2^{2k+p-1}},$$

(a) $$\frac{d}{dx}[x^p J_p(x)] = x^p J_{p-1}(x), \qquad -p \notin N,$$

and for $-p \notin Z^+$

$$\frac{d}{dx}[x^{-p} J_p(x)] = \frac{d}{dx} \sum_{k=0}^{\infty} \frac{(-1)^k}{k! \Gamma(k+p+1)} \frac{x^{2k}}{2^{2k+p}}$$

$$= \sum_{k=1}^{\infty} \frac{(-1)^k 2k}{k! \Gamma(k+p+1)} \frac{x^{2k-1}}{2^{2k+p}}$$

$$= x^{-p} \sum_{k=1}^{\infty} \frac{(-1)^k}{(k-1)! \Gamma(k+p+1)} \frac{x^{2k+p-1}}{2^{2k+p-1}}$$

$$= x^{-p} \sum_{k=0}^{\infty} \frac{(-1)^{k+1}}{k! \Gamma(k+p+2)} \frac{x^{2k+p+1}}{2^{2k+p+1}} = -x^{-p} J_{p+1}(x),$$

(b) $$\frac{d}{dx}[x^{-p} J_p(x)] = -x^{-p} J_{p+1}(x), \qquad -p \notin Z^+.$$

Formulas (b) and (1) give

(c) $$\frac{d}{dx} J_0(x) = -J_1(x) = J_{-1}(x).$$

Hence formula (a) also holds for $p = 0$. Suppose that p is a negative integer. Set $n = -p$. Formulas (a) and (b) then give

$$\frac{d}{dx}[x^n J_n(x)] = x^n J_{n-1}(x)$$

and

$$\frac{d}{dx}[x^{-n} J_n(x)] = -x^{-n} J_{n+1}(x).$$

Thus by (1)

$$\frac{d}{dx}[x^n J_{-n}(x)] = -x^n J_{-n+1}(x)$$

and

$$\frac{d}{dx}[x^{-n}J_{-n}(x)] = x^{-n}J_{-n-1}(x).$$

Hence formulas (a) and (b) also hold when p is a negative integer. In summary, formulas (a) and (b) are valid for every real number p. They give for $p \in R$

$$x^p J_p'(x) + px^{p-1}J_p(x) = x^p J_{p-1}(x),$$
$$x^{-p}J_p'(x) - px^{-p-1}J_p(x) = -x^{-p}J_{p+1}(x),$$

and further

$$J_p'(x) + \frac{p}{x}J_p(x) = J_{p-1}(x), \quad x > 0, \quad p \in R, \tag{4}$$

$$J_p'(x) - \frac{p}{x}J_p(x) = -J_{p+1}(x), \quad x > 0, \quad p \in R. \tag{5}$$

Subtraction and addition member by member of (4) and (5) give (2) and (3). Thus the theorem is proved.

As by-products of the proof we have formulas (4) and (5), and the three formulas

$$\frac{d}{dx}[x^p J_p(x)] = x^p J_{p-1}(x), \quad x > 0, \quad p \in R, \tag{6}$$

$$\frac{d}{dx}[x^{-p}J_p(x)] = -x^{-p}J_{p+1}(x), \quad x > 0, \quad p \in R, \tag{7}$$

$$\frac{d}{dx}J_0(x) = -J_1(x). \tag{8}$$

6.4.3 Example
Set $p = \frac{1}{2}$ in (2), and use formulas (4) and (5) of Section 6.3.2. We get

$$J_{3/2}(x) = \frac{1}{x}J_{\frac{1}{2}}(x) - J_{-\frac{1}{2}}(x) = \sqrt{\frac{2}{\pi x}}\left(\frac{\sin x}{x} - \cos x\right), \quad x > 0.$$

It is analogously seen that $J_p(x)$ is an elementary function for $p = m + \frac{1}{2}$ where m is an arbitrary integer.

EXERCISES

618. It is known that $J_0(2) \approx 0.224$ and $J_1(2) \approx 0.577$. Compute approximate values for
    a) $J_2(2)$, b) $J_1'(2)$, c) $J_2'(2)$.

619. Find $J_{-3/2}(x)$ in terms of elementary functions.

620. Find $J_{5/2}(x)$ in terms of elementary functions.

*621. a) Show that the function $xJ_0(x)$ is a solution of the differential equation

$$y'' + y = -J_1(x), \quad x > 0.$$

b) Use the result in (a) to prove that

$$\int_0^x \cos(x - t)J_0(t)\, dt = x\, J_0(x), \qquad x > 0.$$

622. Show that $\displaystyle\int_0^x tJ_0(t)\, dt = xJ_1(x), \qquad x > 0.$

623. Suppose that p is a real number and that m is a positive integer. Show that

$$2^m \frac{d^m}{dx^m} J_p(x) = \sum_{k=0}^m (-1)^{m-k}\binom{m}{k} J_{p+m-2k}(x), \qquad x > 0.$$

624. Show that $\displaystyle\sum_{n=0}^\infty J_{2n+1}(x) = \tfrac{1}{2}\int_0^x J_0(t)\, dt, \qquad x > 0.$

Hint. Formula (3) gives

$$J_0(x) = 2J_1'(x) + J_2(x) = 2J_1'(x) + 2J_3'(x) + J_4(x), \text{ etc.}$$

Use the estimate in Exercise 615.

6.5 ESTIMATION OF BESSEL FUNCTIONS FOR LARGE VALUES OF x. THE ZEROS OF THE BESSEL FUNCTIONS

Fig. 6.2 and formulas (4) and (5) of Section 6.3.2 indicate that a theorem such as the following should hold.

6.5.1   Theorem. *Suppose that p is a real number. Then there exist a positive number r_p and a real number θ_p, such that the Bessel function of order p has the property*:

$$\sqrt{x}\, J_p(x) - r_p \sin(x + \theta_p) \to 0 \qquad as \qquad x \to \infty. \tag{1}$$

Proof. Suppose $x > 0$. We have by Lemma 6.2.2, Definition 6.2.3, and Definition 6.4.1

$$x^2 J_p''(x) + xJ_p'(x) + (x^2 - p^2)J_p(x) = 0.$$

The substitution $\sqrt{x}\,J_p(x){=}z(x)$ gives (Exercise 625)

(a) $$z'' + z - \frac{p^2 - \tfrac{1}{4}}{x^2}\, z = 0.$$

Consider the function $r(x) = [z^2(x) + z'^2(x)]^{\frac{1}{2}}$. Suppose that $r(x)$ has a zero x_0. Then $z(x_0) = z'(x_0) = 0$, and Lemma 5.4.2 shows that $z(x) = 0$ for every $x > 0$. This, however, is impossible. Then $r(x)$ takes only positive values. Further, $r(x)$ is a continuous function. Then there is a continuous function $\theta(x)$ such that

(b) $$z(x) = r(x) \sin \theta(x) \qquad \text{and} \qquad z'(x) = r(x) \cos \theta(x).$$

Substitution into (a) gives

(c) $$r' \cos \theta - r(\sin \theta)\theta' + r \sin \theta - \frac{p^2 - \frac{1}{4}}{x^2} r \sin \theta = 0.$$

Further it follows from (b) that

(d) $$r' \sin \theta + r(\cos \theta)\theta' - r \cos \theta = 0.$$

Elimination of θ' and r' between (c) and (d) gives

$$r' = \frac{p^2 - \frac{1}{4}}{x^2} r \sin \theta \cos \theta,$$

$$\theta' = 1 - \frac{p^2 - \frac{1}{4}}{x^2} \sin^2\theta.$$

Then (we write exp a for e^a)

$$r(x) = r(1) \exp \int_1^x \frac{p^2 - \frac{1}{4}}{t^2} \sin \theta(t) \cos \theta(t)\, dt.$$

As $\left|\sin \theta(t) \cos \theta(t)\right| \leqslant \frac{1}{2}$ for every t, it follows that $\lim_{x \to \infty} r(x) = r_p$ exists and is positive. Further

$$\theta(x) = \theta(1) + \int_1^x \left[1 - \frac{p^2 - \frac{1}{4}}{t^2}\right] \sin^2\theta(t)\, dt.$$

Then also $\lim_{x \to \infty} [\theta(x) - x] = \theta_p$ exists. Thus the theorem is proved.

We can now give a theorem on the zeros of the Bessel functions. It should be observed that the numbers $j_1, j_2, \ldots$ in the statement of the theorem depend on p.

6.5.2 Theorem. *Suppose that p is a real number. Then the Bessel function of order p has infinitely many positive zeros. These are all simple zeros. Further they can be arranged in an increasing sequence $j_1 < j_2 < \cdots < j_n < \cdots$, such that $j_n \to \infty$ as $n \to \infty$.*

Proof. Let r_p and θ_p have the same meaning as in Theorem 6.5.1. Suppose that $0 < \varepsilon < r_p$. Theorem 6.5.1 shows that, for sufficiently large values of x, the curve $y = \sqrt{x} J_p(x)$ is situated in the "channel" between the curves $y = r_p \sin(x + \theta_p) + \varepsilon$ and $y = r_p \sin(x + \theta_p) - \varepsilon$. Hence $J_p(x)$ has infinitely many zeros. Further (cf. the proof of Theorem 6.5.1) the existence of a multiple zero would have the absurd consequence that $J_p(x)$ is zero for each $x > 0$. Hence every zero is simple. Finally every zero is isolated (cf. [1], p. 127), i.e. each zero has a neighborhood that does not contain another zero. Thus the last claim holds also.

EXERCISES

625. Deduce the differential equation (a).

*626. Find the numbers r_p and θ_p of Theorem 6.5.1 for $p = \frac{1}{2}, p = -\frac{1}{2}, p = \frac{3}{2}$.

627. Suppose that $p \geqslant 0$. Show that the Bessel function $J_p(x)$ is bounded.

6.6 BESSEL SERIES

Suppose that $p \geqslant 0$ or that p is a negative integer. Then the Bessel function $J_p(x)$ is bounded (see Exercise 627 and Definition 6.4.1). Further it is integrable on every finite subinterval of $(0, \infty)$. Now suppose that $-1 < p < 0$. Then $J_p(x)$ is unbounded on every interval of the form $(0, x_0)$. It is, however, integrable on every finite subinterval of $(0, \infty)$. For the remaining values of p, $J_p(x)$ is not integrable on any interval of the form $(0, x_0)$.

Therefore, in this section we assume that $p \geqslant 0$. (A similar but more complicated discussion can be given for $-1 < p < 0$.) The existence of the zeros referred to in the following theorem is proved in Theorem 6.5.2. See also Definition 2.6.4.

6.6.1 Theorem. *Suppose that p is a nonnegative real number, that $J_p(x)$ is the Bessel function of order p, and that j_n, $n \in Z^+$, are the positive zeros of $J_p(x)$. Then the system*

$$J_p(j_n x), \qquad n \in Z^+, \quad x \in (0, 1), \tag{1}$$

is an orthogonal system with the weight function $w(x) = x$.

Proof. Suppose that α and β are two positive zeros of $J_p(x)$. Set $u(x) = J_p(\alpha x)$, $v(x) = J_p(\beta x)$, and $y = \alpha x$. Then

$$y^2 J_p''(y) + y J_p'(y) + (y^2 - p^2) J_p(y) = 0,$$

$$\alpha^2 x^2 J_p''(\alpha x) + \alpha x J_p'(\alpha x) + (\alpha^2 x^2 - p^2) J_p(\alpha x) = 0,$$

(a) $\qquad\qquad x^2 u'' + x u' + (\alpha^2 x^2 - p^2) u = 0,$

and similarly

(b) $\qquad\qquad x^2 v'' + x v' + (\beta^2 x^2 - p^2) v = 0.$

Elimination of p^2 between (a) and (b) gives

$$\frac{d}{dx} [x(u'v - uv')] + (\alpha^2 - \beta^2) x u v = 0.$$

Integration on the interval $(0, 1)$ yields

$$0 = (\alpha^2 - \beta^2) \int_0^1 x u(x) v(x) \, dx = (\alpha^2 - \beta^2) \int_0^1 x J_p(\alpha x) J_p(\beta x) \, dx.$$

This proves the assertion.

6.6.2 Problem

Find the norms of the functions in the orthogonal system (1).

Solution. It follows from (a) that

$$2x^2 u'u'' + 2xu'^2 + 2(\alpha^2 x^2 - p^2)uu' = 0,$$

$$\frac{d}{dx}[x^2 u'^2 + (\alpha^2 x^2 - p^2)u^2] - 2\alpha^2 xu^2 = 0.$$

Integration on the interval $(0, 1)$ gives

$$u'^2(1) - 2\alpha^2 \int_0^1 xu^2(x)\, dx = 0, \qquad J_p'^2(\alpha) - 2\int_0^1 xJ_p^2(\alpha x)\, dx = 0.$$

We have by formula (10) in Definition 2.6.4

$$\| J_p(j_n x) \|^2 = \int_0^1 xJ_p^2(j_n x)\, dx, \qquad n \in Z^+.$$

Formula (5) in Theorem 6.4.2 then shows that

$$\| J_p(j_n x) \|^2 = \tfrac{1}{2}J_{p+1}^2(j_n), \qquad n \in Z^+. \tag{2}$$

6.6.3 Definitions

Suppose that $f(x)$ is a complex-valued function and that the product $xf(x)$ is integrable on the interval $(0, 1)$. Then, by Definition 2.6.4, $f(x)$ has Fourier coefficients c_n with respect to the system (1):

$$c_n = 2[J_{p+1}(j_n)]^{-2} \int_0^1 xf(x)J_p(j_n x)\, dx, \qquad n \in Z^+, \tag{3}$$

and a Fourier series with respect to the system (1):

$$f(x) \sim \sum_{n=1}^{\infty} c_n J_p(j_n x). \tag{4}$$

The numbers (3) are called the *Bessel coefficients of order p* of $f(x)$, and the series in (4) is called the *Bessel series of order p* of $f(x)$ (also the Fourier–Bessel series of the first kind of order p).

The Parseval property and the notion of completeness are defined for an orthogonal system with a weight function in a manner analogous to that of Definition 2.3.4. The system (1) has the Parseval property (see [18], p. 204). It follows that the system (1) is complete, and that, if the function $f(x)$ is square integrable on $(0, 1)$, the series in (4) converges in the mean to $f(x)$. Theorems on pointwise convergence of the series in (4) are also known (see [17]).

EXERCISES

628. Sketch the curves $y = J_0(j_n x)$, $0 < x < 1$, $n = 1, 2, 3$.
    Hint. Use Fig. 6.2.

*629. Suppose that $p \geqslant 0$. Expand the function $f(x) = x^p$, $0 < x < 1$, in a Bessel series of order p.
    Hint. Use formula (6) of Theorem 6.4.2 to evaluate the right member of

$$\int_0^1 x^{1+p} J_p(j_n x)\, dx = j_n^{-2-p} \int_0^{j_n} x^{1+p} J_p(x)\, dx.$$

630. Expand the function $f(x) = x^2$, $0 < x < 1$, in a Bessel series of order zero.
    Hint. Use the same device as in the previous exercise; also use formula (8) of Theorem 6.4.2.

6.7 THE GENERATING FUNCTION OF THE BESSEL FUNCTIONS OF INTEGRAL ORDER

6.7.1 Theorem (giving the generating function of the Bessel functions of integral order). *Suppose that x is a positive number and t a nonzero complex number. Then*

$$e^{x(t-1/t)/2} = \sum_{n=-\infty}^{\infty} J_n(x)t^n, \qquad x > 0, \quad t \in C, \quad t \neq 0, \tag{1}$$

where $J_n(x)$ is the value of the Bessel function of order n at the point x.

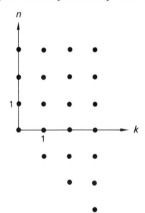

Figure 6.3

Proof. The substitution $m = k + n$ gives the last equality sign in the following computation (the dots in Fig. 6.3 indicate the pairs (k, n) in the last member):

$$e^{x(t-1/t)/2} = e^{xt/2} e^{-x/(2t)}$$

$$= \sum_{m=0}^{\infty} \frac{1}{m!} \left(\frac{xt}{2}\right)^m \sum_{k=0}^{\infty} \frac{1}{k!} \left(-\frac{x}{2t}\right)^k$$

$$= \sum_{k=0}^{\infty} \sum_{m=0}^{\infty} \frac{(-1)^k}{k!m!} \left(\frac{x}{2}\right)^{m+k} t^{m-k}$$

$$= \sum_{k=0}^{\infty} \sum_{n=-k}^{\infty} \frac{(-1)^k}{k!(k+n)!} \left(\frac{x}{2}\right)^{2k+n} t^n.$$

First suppose that $n \geqslant 0$. Then t^n in the last member has the coefficient

$$\sum_{k=0}^{\infty} \frac{(-1)^k}{k!(k+n)!} \left(\frac{x}{2}\right)^{2k+n}.$$

Hence the assertion holds for $n \geqslant 0$. Now suppose that $n < 0$, and set $m = -n$. Then t^n has the coefficient (Definition 6.4.1 gives the last equality sign):

$$\sum_{k=-n}^{\infty} \frac{(-1)^k}{k!(k+n)!} \left(\frac{x}{2}\right)^{2k+n} = \sum_{k=m}^{\infty} \frac{(-1)^k}{k!(k-m)!} \left(\frac{x}{2}\right)^{2k-m} = \sum_{k=0}^{\infty} \frac{(-1)^{k+m}}{(k+m)!k!} \left(\frac{x}{2}\right)^{2k+m}$$

$$= (-1)^m J_m(x) = (-1)^n J_{-n}(x) = J_n(x).$$

Hence the assertion holds for $n < 0$ also.

6.7.2   Corollary. *Suppose that n is an integer. Then the Bessel function of order n satisfies the equation*

$$J_n(x) = \frac{1}{\pi} \int_0^{\pi} \cos(x \sin \theta - n\theta) \, d\theta, \qquad x > 0, \quad n \in Z. \tag{2}$$

Proof. The formula for the nth coefficient of a Laurent series (see [1], p. 183) and the substitution $z = e^{i\theta}$ give

$$J_n(x) = \frac{1}{2\pi i} \int_{|z|=1} e^{x(z-1/z)/2} \frac{dz}{z^{n+1}} = \frac{1}{2\pi} \int_{-\pi}^{\pi} e^{i(x \sin \theta - n\theta)} d\theta$$

$$= \frac{1}{2\pi} \int_{-\pi}^{\pi} \cos(x \sin \theta - n\theta) \, d\theta,$$

and (2) follows.

EXERCISES

631. Suppose that n is an integer. Show that $|J_n(x)| < 1$ for $x > 0$.

*632. Suppose that x is a positive number and θ a real number. Show that

$$\cos (x \sin \theta) = J_0(x) + 2 \sum_{n=1}^{\infty} J_{2n}(x) \cos 2n\theta,$$

$$\sin (x \sin \theta) = 2 \sum_{n=1}^{\infty} J_{2n-1}(x) \sin (2n - 1)\theta.$$

633. Suppose that x is a positive number. Show that

$$\cos x = J_0(x) + 2 \sum_{n=1}^{\infty} (-1)^n J_{2n}(x),$$

$$\sin x = 2 \sum_{n=1}^{\infty} (-1)^{n-1} J_{2n-1}(x).$$

Hint. Use the result in Exercise 632.

634. Suppose that x and y are positive numbers and n an integer. Show that

$$J_n(x + y) = \sum_{m=-\infty}^{\infty} J_m(x) J_{n-m}(y).$$

635. Show that $J_0^2(x) + 2\sum_{n=1}^{\infty} J_n^2(x) = 1$ for $x > 0$, and that $|J_n(x)| \leqslant 2^{-\frac{1}{2}}$ for $x > 0$ and $n \in Z^+$.
Hint. Express the constant term in the Laurent series in t of the function $\exp [x(t - 1/t)/2] \exp [-x(t - 1/t)/2]$ in terms of the coefficients of the Laurent series of the factors.

6.8 NEUMANN FUNCTIONS

6.8.1 Definitions

The limit

$$\gamma = \lim_{n \to \infty} (1 + \frac{1}{2} + \cdots + \frac{1}{n} - \log n) \tag{1}$$

is called *Euler's constant*. Formula (1) shows that γ is equal to the sum of the areas of an infinite sequence of roughly triangular domains, determined by the curve $y = (x + 1)^{-1}$, $x \geqslant 0$, as indicated in Fig. 6.4. Hence γ is a little larger than $\frac{1}{2}$. It can be shown that $\gamma \approx 0.5772$.

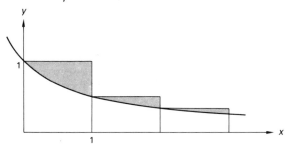

Figure 6.4

Suppose that n is a nonnegative integer. Definition of and notation for the *Neumann function of order n* (also called the Bessel function of second kind of order n) are given by the equations

$$Y_0(x) = -\frac{2}{\pi} \sum_{k=1}^{\infty} \frac{(-1)^k h_k}{(k!)^2} \left(\frac{x}{2}\right)^{2k} + \frac{2}{\pi} J_0(x) \left(\log \frac{x}{2} + \gamma\right), \qquad (2)$$

$$Y_n(x) = -\frac{1}{\pi} \sum_{k=0}^{n-1} \frac{(n-k-1)!}{k!} \left(\frac{x}{2}\right)^{2k-n} - \frac{h_n}{\pi(n!)} \left(\frac{x}{2}\right)^n$$

$$-\frac{1}{\pi} \sum_{k=1}^{\infty} \frac{(-1)^k (h_k + h_{k+n})}{k!(k+n)!} \left(\frac{x}{2}\right)^{2k+n} + \frac{2}{\pi} J_n(x) \left(\log \frac{x}{2} + \gamma\right), \quad n \in Z^+, \qquad (3)$$

where

$$h_k = 1 + \frac{1}{2} + \cdots + \frac{1}{k},$$

and where γ is Euler's constant. Fig. 6.5 shows the graphs of some Neumann functions.

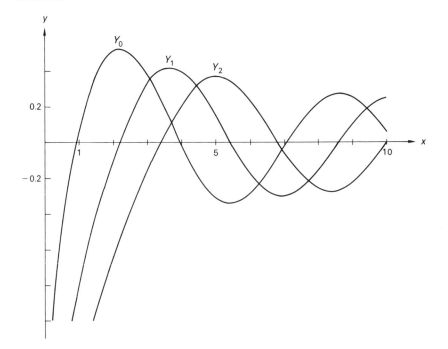

Figure 6.5

6.8.2 Lemma. *Suppose that n is a nonnegative integer. Then the Neumann function of order n is a solution of the Bessel differential equation of order n.*

The lemma can be proved by substitution of (2) and (3) respectively in the left member of the differential equation; the computational work is tedious (Exercise 640). Formulas (2) and (3) are complicated and give the impression of being artificial. There is, however, a "natural" deduction of these formulas, including a proof of the lemma (see [18], pp. 54–55; the computations in this deduction are also long).

We can now give a theorem that covers the case left open in Theorem 6.2.4.

6.8.3 Theorem. *Suppose that n is a nonnegative integer. Then the Bessel differential equation of order n has the general solution*

$$y = c_1 J_n(x) + c_2 Y_n(x), \qquad x > 0, \quad n \in N. \tag{4}$$

Proof. Lemmas 6.2.2 and 6.8.2 show that the functions $J_n(x)$ and $Y_n(x)$ are solutions. They are further linearly independent, for $Y_n(x) \to -\infty$ as $x \to 0$. Hence, by the result in Exercise 608, the assertion holds.

6.8.4 Corollary. *Suppose that n is a nonnegative integer. Then there exist a positive number r_n and a real number θ_n, such that the Neumann function of order n has the property*

$$\sqrt{x}\, Y_n(x) - r_n \sin(x + \theta_n) \to 0 \qquad as \quad x \to \infty. \tag{5}$$

Proof. In the proof of Theorem 6.5.1 we used no other property of the function $J_p(x)$ than that it satisfies the Bessel differential equation of order p. The same proof then gives (5).

EXERCISES

636. Solve the differential equation $x^2 y'' + xy' + (x^2 - 9)y = 0$, $x > 0$.

637. Solve the differential equations of Exercises 611, 612, and 613.

638. Find a bounded solution $y(x)$ of the differential equation $x^2 y'' + xy' + (x^2 - 4)y = 0$, $x > 0$, such that a) $y(2) = 3$, b) $y(j_1) = 3$, where j_1 is the smallest positive zero of the Bessel function $J_2(x)$.

*639. State and prove a theorem for the Neumann functions analogous to Theorem 6.5.2.

640. Show that the function (2) is a solution of the Bessel differential equation of order zero.

641. Formulate a remark on the Bessel functions, analogous to Remark 3.3.2. In particular:
    a) Give a differential equation similar to (2a) in Definition 2.6.2.
    b) Give boundary conditions for $x = 0$ and $x = 1$.
    c) Give a sequence of eigenvalues.
    d) Give corresponding eigenfunctions.

CHAPTER 7

PARTIAL DIFFERENTIAL EQUATIONS
OF FIRST ORDER

In Chapters 7 and 8 all functions are real-valued functions of one or more real variables, unless the context indicates otherwise.

A region Ω in a plane is a nonempty connected open set in the plane. That Ω is open means that, for every point P in Ω, there is a circular disk with center P that is a subset of Ω. That Ω is connected means that any two points in Ω can be joined by a polygonal line in Ω.

7.1 INTRODUCTION

7.1.1 Definitions

A real-valued function $f(x)$ belongs to the *class* $C^1(a, b)$ if its domain is the interval (a, b) and if it has a continuous first derivative on (a, b). (The superscript 1 refers to the order of the mentioned derivative.) Certain other classes are defined analogously, e.g.

$$C^1[a, b], \qquad C^2(a, b), \qquad C^1(\Omega);$$

in the last example, Ω is a region in the xy-plane, and the functions $u(x, y)$ of the class have continuous partial derivatives of first order, $\partial u/\partial x$ and $\partial u/\partial y$, in Ω.

7.1.2 Definitions

Suppose that $F(x, y, u, p, q)$ is a real-valued function of five variables, whose domain is the set of points in the space R^5 (the space of ordered 5-tuples of real numbers) such that (x, y) belongs to a given region Ω in the xy-plane. A *first-order partial differential equation* in two independent variables x and y, *given* in the region Ω, is an equation of the form

$$F\left(x, y, u, \frac{\partial u}{\partial x}, \frac{\partial u}{\partial y}\right) = 0, \qquad (x, y) \in \Omega, \tag{1}$$

where F satisfies the mentioned conditions. A *solution* of (1) is a function $u(x, y)$ of the class $C^1(\Omega)$ that together with its partial derivatives satisfies equation (1) for all $(x, y) \in \Omega$. Its *general solution* is the set of all its solutions. To *solve* (1) is to determine its general solution. If $u(x, y)$ is a solution of (1), the surface $z = u(x, y)$ in the xyz-space is a *solution surface* of (1).

We use the letter u in two meanings: u denotes a dependent variable or a function. The letter z is later used analogously. This is established usage and causes no difficulty: the meaning will always be clear from the context.

We shall sometimes denote partial derivatives by primes and subscripts. For example we often write

$$u'_x, u''_{xx}, u''_{xy} \quad \text{for} \quad \frac{\partial u}{\partial x}, \frac{\partial^2 u}{\partial x^2}, \frac{\partial^2 u}{\partial x \partial y}.$$

(The author has found the notation u''_{xx} to be clearer than the familiar notation u_{xx} for the second partial derivative $\partial^2 u/\partial x^2$.)

Observe that, in the above definition of a solution, it is required that the partial derivatives be continuous. If this restriction were omitted, a discontinuous function might be a solution of a partial differential equation (cf. Exercise 703); such a phenomenon seems awkward.

It is of course possible to study other first-order partial differential equations than (1), namely differential equations with more than two independent variables and such equations with more than one dependent variable (systems of differential equations). We make no such study in this text.

7.1.3 Definitions

The differential equation (1) is *homogeneous* if it is of the form

$$a(x, y)\frac{\partial u}{\partial x} + b(x, y)\frac{\partial u}{\partial y} = 0, \tag{2}$$

it is *linear* if it is of the form

$$a(x, y)\frac{\partial u}{\partial x} + b(x, y)\frac{\partial u}{\partial y} + c(x, y)u = d(x, y), \tag{3}$$

and it is *quasilinear* if it is of the form

$$a(x, y, u)\frac{\partial u}{\partial x} + b(x, y, u)\frac{\partial u}{\partial y} = c(x, y, u); \tag{4}$$

here a, b, c, d denote given functions. Here obviously (2) is a special case of (3), and (3) is a special case of (4). In this chapter we shall restrict our study to differential equations of the forms (2), (3) and (4).

7.1.4 Convention

In Definitions 7.1.2 a region Ω is given. We shall see in a couple of examples that such a specification is essential. In these examples we shall need the notion of an *arbitrary function* $A(x)$; such a function can be arbitrarily chosen in the class $C^1(R)$. We reserve the notation $A(x)$ for arbitrary functions.

7.1.5 Example

Solve the differential equation

$$\frac{\partial u}{\partial y} = 0, \tag{5}$$

given in the entire xy-plane.

Solution. **1.** Suppose that $A(x)$ is an arbitrary function. Then

$$z = A(x) \tag{6}$$

is a solution of (5).

2. Suppose that $z = u(x, y)$ is a solution of (5). Set $A(x) = u(x, 0)$. Then $A(x) = u(x, y)$ for all x and y.

3. Summing up, (6) is the general solution (5).

7.1.6 Example

Consider the differential equation (5) in the region Ω that consists of the xy-plane except the origin and the positive x-axis. Does (6) give the general solution?

Solution. The function

$$u(x, y) = \begin{cases} x^2 & \text{if} \quad x > 0 \quad \text{and} \quad y > 0 \\ 0 & \text{otherwise in } \Omega \end{cases}$$

is a solution (see Figure 7.1; it shows the surface $z = u(x, y)$ for $|x| \leqslant 1$ and $|y| \leqslant 1$). Hence the answer of the question is in the negative.

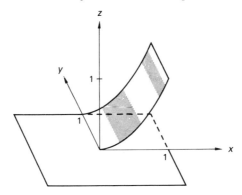

Figure 7.1

In Examples 7.1.5 and 7.1.6 we can make some observations. In the general solution of a partial differential equation arbitrary functions can be met with. The restrictions to a region Ω_1 of the functions in the general solution of a differential equation, given in a region Ω that has Ω_1 as a subregion, need not form the general solution of the same differential equation given in Ω_1. The idea in Example 7.1.6 can be used to construct regions Ω where it would be a complicated task to describe the general solution of (5), given in Ω. These observations induce us to introduce a convention and a definition.

7.1.7 Convention

When we consider a partial differential equation without indicating a region for it, we suppose that it is given in the largest region where the conditions mentioned in the context of (1) are fulfilled. For example, if the differential equations

$$2x\frac{\partial u}{\partial x} + 3y\frac{\partial u}{\partial y} = 0 \quad \text{and} \quad \frac{\partial u}{\partial x} + \sqrt{y}\,\frac{\partial u}{\partial y} = 0$$

are given without corresponding regions, they are supposed to be given in the entire xy-plane and in the half-plane $y > 0$ respectively.

7.1.8 Definition

By a *family of solutions* of a differential equation (1) we shall understand an infinite set of solutions given by

$$z = f(x, y, A), \quad \text{where} \quad A = (g(x, y)), \tag{7}$$

where f and g are given functions and A is an arbitrary function. Example 7.1.6 shows that a family of solutions need not be the general solution.

EXERCISES

701. Solve the differential equations

a) $\dfrac{\partial z}{\partial x} = f(x)$,   b) $\dfrac{\partial z}{\partial x} = f(y)$,

where f is a continuous function, defined on an open interval.

*702. Find the general solution of the differential equation (5), given in the region Ω of Example 7.1.6.

703. Set $u(x, y) = xy/(x^2 + y^2)$ for $(x, y) \neq (0, 0)$ and $u(0, 0) = 0$. Show that the function $u(x, y)$ is discontinuous at the origin, and that it satisfies the equation

$$x\frac{\partial u}{\partial x} + y\frac{\partial u}{\partial y} = 0$$

in the entire xy-plane.

7.2 THE DIFFERENTIAL EQUATION OF A FAMILY OF SURFACES

In this section we give examples showing that it is sometimes possible to associate a first-order partial differential equation with a given set of functions of two variables.

7.2.1 Definition

Suppose that a and b are parameters, that f is a given function of four variables, that Ω is a region in the xy-plane, and that the equation

$$z = f(x, y, a, b), \tag{1}$$

for given values of a and b, always denotes a surface in the xyz-space whose projection on the xy-plane is Ω. It may be possible to eliminate a and b between (1) and its partial derivatives

$$\frac{\partial z}{\partial x} = f'_x(x, y, a, b) \quad \text{and} \quad \frac{\partial z}{\partial y} = f'_y(x, y, a, b).$$

If so, we get a partial differential equation

$$F\left(x, y, z, \frac{\partial z}{\partial x}, \frac{\partial z}{\partial y}\right) = 0, \qquad (x, y) \in \Omega, \tag{2}$$

that is satisfied by each function of the form (1). Then (2) is called *the differential equation of the family of surfaces* (1).

7.2.2 Example

Find the differential equation of the family of planes that intersect the z-axis at the origin.

Solution. The equation (1) is here

$$z = ax + by.$$

Partial differentiations give

$$\frac{\partial z}{\partial x} = a \qquad \text{and} \qquad \frac{\partial z}{\partial y} = b.$$

Elimination of a and b gives the differential equation asked for:

$$x \frac{\partial z}{\partial x} + y \frac{\partial z}{\partial y} - z = 0.$$

7.2.3 Definition

Suppose that Ω is a region in the xy-plane, that $f(x, y)$ is a function with domain Ω, and that A is an arbitrary function. Then, for given A, the equation

$$z = A(f(x, y)) \tag{3}$$

denotes a surface in the xyz-space whose projection on the xy-plane is Ω. It may be possible to eliminate A between the first-order partial derivatives of (3). If so, we get a partial differential equation (2), given in Ω, that is satisfied by each function of the form (3). Then (2) is called *the differential equation of the family of surfaces* (3).

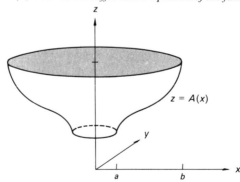

Figure 7.2

7.2.4 Example

Suppose that $0 \leqslant a < b \leqslant \infty$ and that A is an arbitrary function with domain (a, b). Let the curve $z = A(x)$ generate a surface by rotation around the z-axis (see Figure 7.2). Find the differential equation of the family of such surfaces of revolution.

Solution. Set $r = [x^2 + y^2]^{1/2}$. The function $f(x, y)$ of (3) here is r with the "annular ring" $a < r < b$ as domain. The surfaces of revolution are given by the equation $z = A(r)$. Partial differentiations give

$$\frac{\partial z}{\partial x} = A'(r)\frac{x}{r} \quad \text{and} \quad \frac{\partial z}{\partial y} = A'(r)\frac{y}{r}.$$

Elimination of $A'(r)$ gives the differential equation asked for:

(a) $$y\frac{\partial z}{\partial x} - x\frac{\partial z}{\partial y} = 0.$$

7.2.5 Remark

Suppose that $z = u(x, y)$ is one of the surfaces of Example 7.2.4, and that $P_0 = (x_0, y_0, z_0)$ is a point on this surface. Then the surface has the normal vector $(u'_x(x_0, y_0), u'_y(x_0, y_0), -1)$ at the point P_0. This vector and the vector (x_0, y_0, z_0) lie in a plane through the z-axis. Then the projections of the two vectors on the xy-plane are parallel. This gives the equation (a).

Analogously the equations of the surfaces in the answers of Exercises 707–709 can be obtained.

EXERCISES

704. Find the differential equations of the families of surfaces:
   a) $z = (x + a)(y + b)$, b) $2z = (ax + y)^2 + b$,
   c) $ax^2 + by^2 + z^2 = 1$, $z > 0$ or $z < 0$,
   where a and b are parameters.

705. Find the differential equations of the families of surfaces:
   a) $z = xy + A(x^2 + y^2)$, b) $z = x + y + A(xy)$,

   c) $z = A\left(\dfrac{xy}{z}\right)$, d) $z = A(x - y)$,

   where A is an arbitrary function.

*706. Find the differential equation of the tangent planes of the paraboloid of revolution $z = x^2 + y^2$.

707. Suppose that L is a given straight line in the xy-plane Ω, there having the equation $ax + by + c = 0$, that the function $u(x, y)$ belongs to the class $C^1(\Omega)$, and that the surface $z = u(x, y)$ is a cylinder with generatrices parallel to L, i.e. if (x_0, y_0, z_0) is any point on the surface, then the whole line $a(x - x_0) + b(y - y_0) = 0$, $z = z_0$, lies in the surface. Find the differential equation of the family of such cylinders.

708. Suppose that the region Ω consists of the xy-plane except the origin O, that the function $u(x, y)$ belongs to the class $C^1(\Omega)$, and that the surface $z = u(x, y)$ is a cone with vertex at O, i.e. if P is any point of the surface, then the ray OP (the point O excluded) lies in the surface. Find the differential equation of the family of such cones.

709. Suppose that the region Ω consists of the xy-plane except the origin O, that the function $u(x, y)$ belongs to the class $C^1(\Omega)$, and that if P is any point of the surface $z = u(x, y)$ and Q the projection of P on the z-axis, then the ray QP (the point Q excluded) lies in the surface. (Such a surface is called a conoid.) Find the differential equation of the family of such conoids.

7.3 HOMOGENEOUS DIFFERENTIAL EQUATIONS

In the following equation (1) we require for practical reasons that the functions $a(x, y)$ and $b(x, y)$ belong to the class $C^1(\Omega)$. Comparison between Lemma 7.3.3 and [18], Section 6, shows that this condition can be considerably weakened (it can be replaced by a so-called Lipschitz condition for each closed subset of Ω).

7.3.1 Introductory remark

Suppose that Ω is a region in the xy-plane, and that the functions $a(x, y)$ and $b(x, y)$ belong to the class $C^1(\Omega)$. Consider the homogeneous differential equation

$$a(x, y)\frac{\partial u}{\partial x} + b(x, y)\frac{\partial u}{\partial y} = 0, \qquad (x, y) \in \Omega. \tag{1}$$

Suppose that the curve

$$\left.\begin{array}{l} x = f(t) \\ y = g(t) \\ z = z_0 \end{array}\right\}, \qquad t \in (a, b), \tag{2}$$

where the functions f and g belong to the class $C^1(a, b)$ and z_0 is a constant, is a level curve on a solution surface $z = u(x, y)$ of (1). (See Figure 7.3, where C is a level curve, where C_0 is the projection of C on the xy-plane, and where the dashed

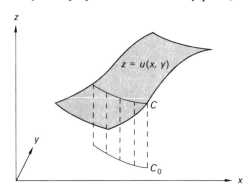

Figure 7.3

segments all have the length z_0.) The composite function $u(f(t), g(t))$ then has the constant value z_0. Hence its derivative vanishes:

$$\frac{\partial u}{\partial x}\frac{dx}{dt} + \frac{\partial u}{\partial y}\frac{dy}{dt} = 0 \qquad \text{for} \quad t \in (a, b),$$

where $dx/dt = f'(t)$ and $dy/dt = g'(t)$. Comparison with (1) makes it probable that there is a close connection between the solution surfaces of (1) and the solution curves of the following system of two ordinary differential equations:

$$\frac{dx}{dt} = a(x, y), \qquad \frac{dy}{dt} = b(x, y), \qquad (x, y) \in \Omega. \tag{3}$$

Such a connection will be stated in Theorem 7.3.4 below. We first give two definitions and a lemma.

7.3.2 Definitions

If a partial differential equation of the form (1) is given, then the equations (3) are called the *characteristic equations* of (1). Each solution curve C_0 of (3) in the xy-plane and each curve C that can be obtained from C_0 by parallel translation in the direction of the z-axis is called a *characteristic curve* of (1). (See the curves C_0 and C in Fig. 7.3.) It may occur that a characteristic curve consists of one single point.

To be able to state a certain uniqueness property below in Lemma 7.3.3, we adopt the convention that if the functions (2) define a characteristic curve, then their restrictions to any proper subinterval of (a, b) do not define a characteristic curve.

7.3.3 Lemma. *Suppose that Ω is a region in the xy-plane, that the functions $a(x, y)$ and $b(x, y)$ belong to the class $C^1(\Omega)$, and that $P_0 = (x_0, y_0, z_0)$ is a point in the xyz-space whose projection on the xy-plane belongs to Ω. Then there passes through P_0 exactly one characteristic curve of the homogeneous differential equation* (1).

For a proof of the lemma, see [18], Section 6.

7.3.4 Theorem. *Suppose that Ω is a region in the xy-plane, and that the functions $a(x, y)$ and $b(x, y)$ belong to the class $C^1(\Omega)$. Then the function $u(x, y)$ is a solution of the homogeneous equation* (1), *if and only if $u(x, y)$ belongs to the class $C^1(\Omega)$ and the surface $z = u(x, y)$ is a union of characteristic curves of* (1).

Proof. 1. Suppose that the function $u(x, y)$ is a solution of (1). Then, by definition, $u(x, y)$ belongs to the class $C^1(\Omega)$. Suppose that $P_0 = (x_0, y_0, z_0)$ is a point on the surface $z = u(x, y)$. By Lemma 7.3.3, exactly one characteristic curve of (1) passes through P_0. Let (2) be this characteristic curve. Suppose that $t = t_0$ corresponds to P_0. Then

$$u(f(t_0), g(t_0)) = z_0,$$

and for all $t \in (a, b)$,

$$\frac{d}{dt} u(f(t), g(t)) = u'_x(f, g)f'(t) + u'_y(f, g)g'(t)$$

$$= u'_x(f, g)a(f, g) + u'_y(f, g)b(f, g) = 0.$$

Hence

$$u(f(t), g(t)) = z_0 \qquad \text{for} \quad t \in (a, b),$$

i.e. the coordinates of each point of the characteristic curve satisfy the equation $z = u(x, y)$. The characteristic curve then is situated on the surface $z = u(x, y)$. It follows that the surface is a union of characteristic curves.

2. Suppose that the function $u(x, y)$ belongs to the class $C^1(\Omega)$ and that the surface $z = u(x, y)$ is a union of characteristic curves. Let (x_0, y_0) be a point in Ω. Set $z_0 = u(x_0, y_0)$ and $P_0 = (x_0, y_0, z_0)$. By Lemma 7.3.3 there passes exactly one characteristic curve of (1) through P_0. At the point P_0, this characteristic curve has the tangent vector $(a(x_0, y_0), b(x_0, y_0), 0)$, and the surface has the normal vector $(u'_x(x_0, y_0), u'_y(x_0, y_0), -1)$. These vectors are perpendicular. Hence

$$a(x_0, y_0)u'_x(x_0, y_0) + b(x_0, y_0)u'_y(x_0, y_0) = 0.$$

It follows that the function $u(x, y)$ is a solution of (1).

The conclusions of parts **1** and **2** prove the theorem.

7.3.5 Theorem. *If the function $u(x, y)$ is a solution of the homogeneous differential equation* (1) *and if A is an arbitrary function, then $A(u(x, y))$ is a family of solutions of* (1).

Proof. We have for all $(x, y) \in \Omega$

$$a(x, y) \frac{\partial}{\partial x} A(u(x, y)) + b(x, y) \frac{\partial}{\partial y} A(u(x, y))$$

$$= A'(u(x, y)) \left[a(x, y) \frac{\partial u}{\partial x} + b(x, y) \frac{\partial u}{\partial y} \right] = 0,$$

for the expression inside the brackets vanishes by assumption. The assertion now follows.

7.3.6 Example

Suppose that $0 \leqslant a < b \leqslant \infty$, and that Ω is the region in the xy-plane determined by the inequalities $a^2 < x^2 + y^2 < b^2$. Solve the differential equation $y(\partial u/\partial x) - x(\partial u/\partial y) = 0$, given in Ω.

Solution. The characteristic equations are

$$\frac{dx}{dt} = y, \qquad \frac{dy}{dt} = -x.$$

They have the general solution

$$x = r \cos(t - \alpha), \qquad y = -r \sin(t - \alpha), \qquad t \in (-\infty, \infty),$$

where r and α are parameters, $a < r < b$. The characteristic curves in the xy-plane are then circles with center at the origin. Every function of the form

(a) $$z = A(x^2 + y^2), \qquad (x, y) \in \Omega,$$

where A is an arbitrary function, belongs to the class $C^1(\Omega)$ and is constant on every characteristic curve in the xy-plane; hence it is a solution by Theorem 7.3.4. (Alternatively, Theorem 7.3.5 can be applied, for it is readily seen that the function $z = x^2 + y^2$ is a solution of the differential equation.) If further the function $u(x, y)$ is a solution, it is of the form (a), for we can set $A(x) = u(\sqrt{x}, 0)$ for $a < x < b$. Hence (a) is the general solution.

7.3.7 Definition
Given a partial differential equation, the problem of finding a solution surface that contains a given curve is called a *Cauchy problem*. We show by an example how some Cauchy problems can be solved.

7.3.8 Example
Find a solution surface for the differential equation $2y(\partial u/\partial x) + x(\partial u/\partial y) = 0$ that contains the parabola $z = x^2$, $y = x$.

Solution. As in Example 7.3.6 it is found that $z = A(x^2 - 2y^2)$ is a family of solutions. The substitutions $z = x^2$ and $y = x$ give $x^2 = A(x^2 - 2x^2)$. This last equation is satisfied if, as function A, we choose $A(u) = -u$. Hence $z = 2y^2 - x^2$ is a solution surface that contains the given parabola.

EXERCISES

710. Solve the differential equation $x(\partial u/\partial x) + y(\partial u/\partial y) = 0$, given in the half-plane $x > 0$.

711. Solve the differential equation $x(\partial u/\partial x) + y^2(\partial u/\partial y) = 0$, given in the quadrant $x > 0, y > 0$.

712. Find the solution surface of the differential equation $2x(\partial u/\partial x) - 3y(\partial u/\partial y) = 0$, given in the half-plane $x > 0$, that contains the straight line $x = 1$, $y = z$.

*713. Solve the differential equation $(\partial u/\partial x) + x(\partial u/\partial y) = 0$, given in the half-plane $y > 0$.

714. In Example 7.3.8, do solution surfaces that contain the given parabola exist, other than $z = 2y^2 - x^2$?

7.4 LINEAR AND QUASILINEAR DIFFERENTIAL EQUATIONS
7.4.1 Remark
The following theorem has a counterpart in the theory for ordinary differential equations (cf. Exercise 522 and [18], Theorem 10.1). For the proof of the theorem it is left to the reader to define the operator (cf. Section 2.6.3)

$$L = a(x, y)\frac{\partial}{\partial x} + b(x, y)\frac{\partial}{\partial y}, \qquad (x, y) \in \Omega,$$

and to show that it is linear, i.e. that if c_1, c_2 are constants and if u_1, u_2 are functions in the class $C^1(\Omega)$, then $L(c_1 u_1 + c_2 u_2) = c_1 L(u_1) + c_2 L(u_2)$.

7.4.2 Theorem. *Suppose that the function $u_p(x, y)$ is a solution of the linear differential equation*

$$a(x, y)\frac{\partial u}{\partial x} + b(x, y)\frac{\partial u}{\partial y} = c(x, y), \qquad (x, y) \in \Omega. \tag{1}$$

If $u_p(x, y)$ is added to each solution of the associated homogeneous equation (for this term see Section 5.4.1):

$$a(x, y)\frac{\partial u}{\partial x} + b(x, y)\frac{\partial u}{\partial y} = 0, \qquad (x, y) \in \Omega, \tag{2}$$

then the general solution of (1) is obtained.

Proof. **1.** Suppose that $u_1(x, y)$ is a function in the general solution of (1). Then

$$L(u_1 - u_p) = L(u_1) - L(u_p) = c - c = 0,$$

i.e. $u_1(x, y) - u_p(x, y)$ is a solution of (2). Hence any solution of (1) can be obtained by adding $u_p(x, y)$ to a certain solution of (2).

2. Suppose that $u_2(x, y)$ is a function in the general solution of (2). Then

$$L(u_2 + u_p) = L(u_2) + L(u_p) = 0 + c = c,$$

i.e. $u_2(x, y) + u_p(x, y)$ is a solution of (1). Hence a certain solution of (1) is obtained by adding $u_p(x, y)$ to any solution of (2).

The conclusions of parts **1** and **2** prove the theorem.

A couple of applications of this theorem are given in Exercises 715 and 716. A somewhat more general variant of the theorem is given in Exercise 717.

Suppose that Ω is a region in the xy-plane. The Cartesian product of Ω and the z-axis, denoted $\Omega \times R$, is the set of all points in the xyz-space whose projections on the xy-plane belong to Ω.

7.4.3 Definitions

Suppose that Ω is a region in the xy-plane and that the functions $a(x, y, z)$, $b(x, y, z)$, $c(x, y, z)$ belong to the class $C^1(\Omega \times R)$. Consider the quasilinear equation

$$a(x, y, u)\frac{\partial u}{\partial x} + b(x, y, u)\frac{\partial u}{\partial y} = c(x, y, u), \qquad (x, y) \in \Omega, \tag{3}$$

and the system of three ordinary differential equations:

$$\frac{dx}{dt} = a(x, y, z), \qquad \frac{dy}{dt} = b(x, y, z), \qquad \frac{dz}{dt} = c(x, y, z), \qquad (x, y, z) \in \Omega \times R. \tag{4}$$

The equations (4) are called the *characteristic equations* of (3). Suppose that

$$x = f(t), \qquad y = g(t), \qquad z = h(t), \qquad t \in (a, b), \tag{5}$$

is a solution curve of (4), and that this curve is as "large" as possible in the sense explained in the last paragraph of Section 7.3.2. Then the curve (5) is called a *characteristic curve* of (3). A curve (5) need not be parallel to the xy-plane (in contrast to the characteristic curves introduced in Definition 7.3.2).

7.4.4 Theorem. *Suppose that Ω is a region in the xy-plane and that the functions $a(x, y, z), b(x, y, z), c(x, y, z)$ belong to the class $C^1(\Omega \times R)$. Then the function $u(x, y)$ is a solution of the quasilinear differential equation (3), if and only if $u(x, y)$ belongs to the class $C^1(\Omega)$ and the surface $z = u(x, y)$ is a union of characteristic curves of (3).*

It is left to the reader to construct a proof of this theorem (Exercise 718), modeled on the proof of Theorem 7.3.4. The computation in that proof is here replaced by:

$$\frac{d}{dt} u(f(t), g(t)) = u'_x(f, g)a(f, g, h) + u'_y(f, g)b(f, g, h) = c(f, g, h),$$

$$u(f(t), g(t)) = z_0 + \int_{t_0}^{t} c(f(\tau), g(\tau), h(\tau))d\tau$$

$$= z_0 + \int_{t_0}^{t} h'(\tau)d\tau = z_0 + h(t) - h(t_0) = h(t).$$

7.4.5 Example

Suppose that the region Ω consists of the xy-plane except the origin O. Solve the differential equation

$$x \frac{\partial z}{\partial x} + y \frac{\partial z}{\partial y} = z, \qquad (x, y) \in \Omega.$$

Solution. The characteristic equations are

$$\frac{dx}{dt} = x, \qquad \frac{dy}{dt} = y, \qquad \frac{dz}{dt} = z.$$

They have the general solution

$$x = c_1 e^t, \qquad y = c_2 e^t, \qquad z = c_3 e^t, \qquad t \in (-\infty, \infty),$$

where c_1, c_2, c_3 are parameters. The statement of the problem shows that the case $c_1 = c_2 = 0$ cannot occur. Thus the characteristic curves to be considered are the rays with O as the end point (the point O excluded) except the positive z-axis and the negative z-axis. Theorem 7.4.4 then shows that the general solution consists of the functions $u(x, y)$ in the class $C^1(\Omega)$ such that the surfaces $z = u(x, y)$ are cones with vertex at O.

EXERCISES

715. Solve the differential equation $x(\partial u/\partial x) + y(\partial u/\partial y) = xy \sin xy$, given in the half-plane $y > 0$.

    Hint. It has a solution of the form $c \cos xy$, where c is a constant.

716. Solve the differential equation $2(\partial u/\partial x) - 3(\partial u/\partial y) = 4x - 9y^2$.

    Hint. It has a solution of the form $f(x) + g(y)$.

*717. For the differential equation (3) of Definition 7.1.3, state and prove a theorem analogous to Theorem 7.4.2.

718. Prove Theorem 7.4.4.

719. Solve the differential equation $(\partial u/\partial x) + (x + y)(\partial u/\partial y) = xu$.

720. Solve the differential equation $x(\partial z/\partial x) + y(\partial z/\partial y) = z$, given in the entire xy-plane (cf. Example 7.4.5).

721. Solve the differential equation $(\partial u/\partial x) - 2(\partial u/\partial y) + 3u = x$.

722. Find the solution surface for the differential equation $(\partial u/\partial x) - 2(\partial u/\partial y) + 3u = 0$ that contains the curve $x = 0$, $z = 1 + \sin y$.

723. Suppose that c is a nonzero constant. Solve the differential equation $y^3(\partial u/\partial x) - xy^2(\partial u/\partial y) = cxu$, given in the half-plane $y > 0$.

    Hint. Introduce new independent variables such that the characteristic curves of the corresponding homogeneous equation $y^3 u_x' - xy^2 u_y' = 0$ become parallel to one of the coordinate axes.

CHAPTER 8

PARTIAL DIFFERENTIAL EQUATIONS OF SECOND ORDER

8.1 PROBLEMS IN PHYSICS LEADING TO PARTIAL DIFFERENTIAL EQUATIONS

In this introductory section (8.1) we study a few of the physical problems that lead to second-order partial differential equations.

The reader will observe in the following that we give the value 1 to certain physical constants. This is to make our arguments look simple, and this value 1 can always be achieved by suitably choosing the units of the physical quantities involved.

8.1.1 Problem

Find a partial differential equation satisfied by a certain function $u(x, t)$ under the following assumptions. A thin uniform rod is rigidly fastened at its end points, the points $x = 0$ and $x = a$ $(a > 0)$ of the x-axis (see Fig. 8.1). With each point of the rod is associated a number x, showing the distance from the origin to the point when the rod is at equilibrium. The rod is undergoing longitudinal vibrations. The deviation of the point x of the rod from its equilibrium position at the time t is $u(x, t)$, where $u(x, t)$ is a function in the class $C^2(\Omega)$, Ω being determined by the inequalities $0 \leqslant x \leqslant a$, $-\infty < t < \infty$. (Here $u(x, t)$ is positive or negative according as the point has deviated to the right or to the left in Fig. 8.1 from its equilibrium position.) The tension in the rod at the point x and the time t is $u'_x(x, t)$. (Here Hooke's law has been applied. The tension is positive in a part of the rod that is stretched, negative in a part that is contracted.) The mass of any segment $(x, x + h)$ of the rod is h. No outer forces are acting on the rod except those keeping its end points fixed. There is no exchange of energy between the rod and its surroundings; there are no transformations of energy in the rod except between kinetic and potential energy.

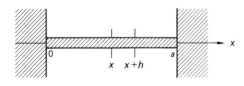

Figure 8.1

131

Solution. The acceleration u_{tt}'' is by assumption a continuous function. Having Newton's law (mass times acceleration equals force) in mind, it seems reasonable to assume that, given a segment $(x, x + h)$ and a time t, there exists a number θ, $0 < \theta < 1$, such that

$$hu_{tt}''(x + \theta h, t) = u_x'(x + h, t) - u_x'(x, t).$$

Divide both members by h, and let h tend to zero. It follows that

$$\frac{\partial^2 u}{\partial t^2} = \frac{\partial^2 u}{\partial x^2}. \tag{1}$$

This equation is called the *wave equation*.

8.1.2 Problem

Find a partial differential equation satisfied by a certain function $u(x, t)$ under the following assumptions. A thin uniform rod has its end points at the points $x = 0$ and $x = a$ $(a > 0)$ of the x-axis (see Fig. 8.1). The temperature at the point x of the rod at the time t is $u(x, t)$, where $u(x, t)$ is a function in the class $C^2(\Omega)$, Ω being determined by the inequalities $0 \leqslant x \leqslant a, 0 \leqslant t < \infty$. The rod does not exchange heat with its surroundings except at its end points. Consider Fig. 8.2. The amount of heat in the segment $(x, x + h)$ of the rod at the time t is $\int_x^{x+h} u(\xi, t) \, d\xi$. The amount of heat that enters the segment $(x, x + h)$ at the point x during the time $(t, t + k)$ is $-\int_t^{t+k} u_x'(x, \tau) \, d\tau$ (that is if this expression has a positive value; otherwise the amount of heat given by this integral with a plus sign leaves the segment).

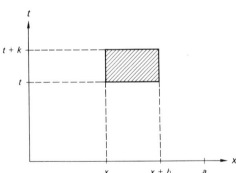

Figure 8.2

Solution. The increase of heat in the segment $(x, x + h)$ during the time $(t, t + k)$ is represented by each member of the equation

$$\int_x^{x+h} u(\xi, t + k) \, d\xi - \int_x^{x+h} u(\xi, t) \, d\xi = \int_t^{t+k} u_x'(x + h, \tau) \, d\tau - \int_t^{t+k} u_x'(x, \tau) \, d\tau.$$

The assumptions then imply the existence of two functions $\varepsilon(h, k)$ and $\eta(h, k)$, both tending to zero as h and k tend to zero, such that

$$hk[u_t'(x, t) + \varepsilon(h, k)] = hk[u_{xx}''(x, t) + \eta(h, k)].$$

Divide both members by hk, and let h and k tend to zero. It follows that

$$\frac{\partial u}{\partial t} = \frac{\partial^2 u}{\partial x^2}. \tag{2}$$

This equation is called the *heat equation*.

The reader may have observed that we gave the interval $-\infty < t < \infty$ in Problem 8.1.1 and the interval $0 \leqslant t < \infty$ in Problem 8.1.2. A reason for this is that the substitution $t = -t'$ in the wave equation leads to the same equation, while this substitution in the heat equation does not give the heat equation. It will be seen in examples in this chapter that, given the state of a rod at the time $t = 0$, it is in general possible to compute its longitudinal vibrations for $t < 0$, and in general impossible to assign a distribution of heat in the rod for $t < 0$. Physicists describe this situation by saying that vibration is a reversible process and that heat conduction is a nonreversible process. In this context the reader might compare the results in Exercise 802 with the second law of thermodynamics.

8.1.3 Problem

Let Ω denote the xy-plane except the origin O. Suppose that a particle, placed at any point P of Ω, is attracted towards the origin by a force, in vector form (p, q), of magnitude r^{-1}, where r is the distance OP (see Fig. 8.3). Let γ be an arc in Ω from P to the fixed point $A = (1, 0)$ with the parametric representation:

$$x = x(t), \qquad y = y(t), \qquad a \leqslant t \leqslant b,$$

where the functions $x(t)$ and $y(t)$ are continuous and piecewise continuously differentiable on the interval $[a, b]$. Let w_γ denote the work performed by the attraction towards O as the particle moves along the curve γ from P to A.

a) Show that w_γ does not depend on the curve γ but only on the point P.

b) Hence $w_\gamma = u(x, y)$, where $u(x, y)$ is a function of the coordinates of P.

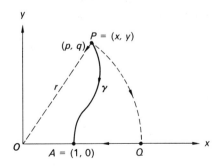

Figure 8.3

The value $u(x, y)$ is called the potential at P. Find a partial differential equation that is satisfied by this function $u(x, y)$.

Solution. a) Let (dx, dy) denote a displacement of the particle, so small that the variation of the force (p, q) can be disregarded. The work then is the scalar product $p\,dx + q\,dy$. Therefore the work w_γ is defined by the line integral

(a)
$$w_\gamma = \int_\gamma p\,dx + q\,dy.$$

There is also a representation by an ordinary integral (cf. [1], pp. 101–106):

(b)
$$w_\gamma = \int_a^b (px'(t) + qy'(t))\,dt.$$

It is readily verified that

(c)
$$p = p(x, y) = -\frac{x}{x^2 + y^2}, \qquad q = q(x, y) = -\frac{y}{x^2 + y^2}.$$

There exists a function $U(x, y)$ such that $p = U'_x$ and $q = U'_y$, e.g. the function $U(x, y) = -\frac{1}{2}\log(x^2 + y^2)$. Then

$$w_\gamma = \int_a^b \left[\frac{d}{dt}U(x(t), y(t))\right] dt = U(x(b), y(b)) - U(x(a), y(a)).$$

Hence w_γ depends only on the point P.

    b) The result just obtained shows that

$$w_\gamma = w_{PQ} + w_{QA},$$

where w_{PQ} and w_{QA} is the work performed when the particle moves along a circular arc PQ and a segment QA of the positive x-axis respectively (see Fig. 8.3). Formulas (b) and (c) show that $w_{PQ} = 0$. Then

$$w_\gamma = w_{QA} = -\int_r^1 \frac{dx}{x} = \log r.$$

Therefore

$$u(x, y) = \tfrac{1}{2}\log(x^2 + y^2),$$
$$u'_x = \frac{x}{x^2 + y^2}, \qquad u''_{xx} = \frac{y^2 - x^2}{(x^2 + y^2)^2},$$
$$u'_y = \frac{y}{x^2 + y^2}, \qquad u''_{yy} = \frac{x^2 - y^2}{(x^2 + y^2)^2}.$$

Hence

$$\frac{\partial^2 u}{\partial x^2} + \frac{\partial^2 u}{\partial y^2} = 0. \tag{3}$$

This equation is called the *Laplace equation*.

EXERCISES

*801. State and solve a problem, analogous to Problem 8.1.1, for transversal vibrations of a string.

802. Consider the function

$$u(x, t) = \frac{1}{\sqrt{4\pi t}} e^{-x^2/(4t)}, \qquad -\infty < x < \infty, \quad t > 0.$$

a) Is $u(x, t)$ a solution of the heat equation?
b) Use formula (9) of Example 4.4.12 to evaluate the integral $\int_{-\infty}^{\infty} u(x, t) \, dx$. Give a physical interpretation of the result.
c) Suppose that $\varepsilon > 0$. Does there exist a solution of the heat equation in the half-plane $t > -\varepsilon$, coinciding with $u(x, t)$ for $t > 0$?

803. Suppose that $P_1, P_2, \ldots, P_n$ and P are $n + 1$ points in the xy-plane, and that a unit mass at P is attracted towards each P_v by a force of magnitude $m_v r_v^{-1}$, where m_v is a positive quantity (a mass placed at P_v) and r_v is the distance PP_v. State and solve a problem analogous to Problem 8.1.3.

804. State and solve a problem, analogous to the problem in Exercise 803, for the xyz-space and forces of magnitude $m_v r_v^{-2}$, $v = 1, 2, \ldots, n$.

8.2 DEFINITIONS
8.2.1 Two independent variables

Suppose that Ω is a region in the xy-plane. A *linear second-order partial differential equation* with *constant coefficients* and with two independent variables x and y, given in Ω, is of the form:

$$a_{11} \frac{\partial^2 u}{\partial x^2} + 2a_{12} \frac{\partial^2 u}{\partial x \partial y} + a_{22} \frac{\partial^2 u}{\partial y^2} + b_1 \frac{\partial u}{\partial x} + b_2 \frac{\partial u}{\partial y} + cu$$

$$= f(x, y), \qquad (x, y) \in \Omega, \tag{1}$$

where the $a_{\mu v}$, b_v, c are given real numbers and at least one of the $a_{\mu v}$ is different from zero, and where f is a given function, defined in Ω. A *solution* of (1) is a function $u(x, y)$ in the class $C^2(\Omega)$ that together with its first-order and second-order partial derivatives satisfies (1) in Ω. The *general solution* of (1) is the set of all its solutions.

If $a_{12} \neq 0$, then (1) can be transformed into an equation of the same form, but without a term with a mixed second derivative, by a change of variables of the form (a rotation of the coordinate axes counterclockwise through an angle θ)

$$x = \alpha \xi - \beta \eta, \qquad y = \beta \xi + \alpha \eta,$$

where

$$\alpha = \cos\theta, \quad \beta = \sin\theta, \quad \cot 2\theta = \frac{a_{11} - a_{22}}{2a_{12}}.$$

The equation obtained can be written

$$\lambda_1 \frac{\partial^2 u}{\partial \xi^2} + \lambda_2 \frac{\partial^2 u}{\partial \eta^2} + \cdots = f(x, y), \tag{2}$$

where λ_1 and λ_2 are constants and where the dots denote terms containing $\partial u/\partial \xi$, $\partial u/\partial \eta$ and u. (The constants λ_1 and λ_2 are the eigenvalues of the matrix

$$\begin{bmatrix} a_{11} & a_{12} \\ a_{12} & a_{22} \end{bmatrix};$$

hence they can readily be found.) If $a_{12} = 0$ in (1), we set $\lambda_1 = a_{11}$ and $\lambda_2 = a_{22}$. The differential equation (1) is said to be *elliptic* if λ_1 and λ_2 have the same sign, *parabolic* if one of λ_1 and λ_2 is zero and *hyperbolic* if λ_1 and λ_2 have different signs. A reason for this classification is that properties and methods of solution for the three types of differential equations show considerable differences.

Now suppose that the differential equation (1) has *variable coefficients* (i.e. at least one among the $a_{\mu\nu}$, b_ν, c is a non-constant function of x and y) and that, for each point (x, y) in Ω, at least one of the $a_{\mu\nu} = a_{\mu\nu}(x, y)$ is different from zero. Let P be a point in Ω. Then the differential equation (1) can be classified as *elliptic*, *parabolic* or *hyperbolic at the point P*. It is left to the reader to define these concepts. An example is given below (the Tricomi equation).

The following list shows name, formula, and type for some special cases of (1):

Laplace equation $\quad \dfrac{\partial^2 u}{\partial x^2} + \dfrac{\partial^2 u}{\partial y^2} = 0 \quad$ elliptic

heat equation $\quad \dfrac{\partial u}{\partial t} - \dfrac{\partial^2 u}{\partial x^2} = 0 \quad$ parabolic

wave equation $\quad \dfrac{\partial^2 u}{\partial t^2} - \dfrac{\partial^2 u}{\partial x^2} = 0 \quad$ hyperbolic

Tricomi equation $\quad \dfrac{\partial^2 u}{\partial x^2} + x \dfrac{\partial^2 u}{\partial t^2} = 0 \quad \begin{cases} \text{elliptic for } x > 0 \\ \text{parabolic for } x = 0 \\ \text{hyperbolic for } x < 0. \end{cases}$

In this list the letter t has been used in places where, in many applications, time is an independent variable.

8.2.2   Three or more independent variables

What has been said above can, with some changes in the formulations, be carried over to the case where we have m independent variables with $m > 2$. The equation (2) then is replaced by

$$\lambda_1 \frac{\partial^2 u}{\partial \xi_1^2} + \cdots + \lambda_m \frac{\partial^2 u}{\partial \xi_m^2} + \cdots = f(\mathbf{x}), \tag{3}$$

where $\mathbf{x} = (x_1, \ldots, x_m)$ is a point in R^m. The given differential equation is *elliptic* at the point $\mathbf{x}$ if all the coefficients $\lambda_1, \ldots, \lambda_m$ have the same sign, *parabolic* if at least one of them is zero, *hyperbolic* if one of them has the opposite sign to the other ones (and "ultrahyperbolic" in the remaining case). (By the Sylvester inertial law (see [5], p. 296), this type is uniquely determined.) Let Δ_m denote the Laplace operator of order m:

$$\Delta_m = \frac{\partial^2}{\partial x_1^2} + \ldots + \frac{\partial^2}{\partial x_m^2}.$$

The following list gives name, formula and type for some partial differential equations (the function $f(\mathbf{x})$ is not identically zero in the region Ω under consideration):

Laplace equation	$\Delta_m u = 0$	elliptic
Poisson equation	$\Delta_m u = f(\mathbf{x})$	elliptic
heat equation	$\frac{\partial u}{\partial x_m} - \Delta_{m-1} u = 0$	parabolic
wave equation	$\frac{\partial^2 u}{\partial x_m^2} - \Delta_{m-1} u = 0$	hyperbolic.

8.2.3 Comment
In the following we restrict the discussion for the most part to the case $m = 2$. Further we let the region Ω be of a simple form, e.g. a half-plane, a quadrant, a circular disk, or the domain $0 < x < \pi, 0 < y < \infty$. In boundary-value problems we require, when nothing else is said, that the solution (or a derivative of the solution) tends to the given boundary value, as (x, y) in Ω approaches any boundary point along the normal of the boundary curve at the point.

8.2.4 Example
The boundary-value problem (Fig. 8.4)

$$\frac{\partial u}{\partial t} - \frac{\partial^2 u}{\partial x^2} = 0, \quad x > 0, \quad t > 0,$$

$$u(x, 0) = 0, \quad x > 0,$$

$$u(0, t) = 0, \quad t > 0,$$

has the solution $u(x, t) = 0$. It also has the solution $v(x, t) = xt^{-3/2}e^{-x^2/(4t)}$, for it can be verified (Exercise 812) that this function satisfies the given differential equation for positive values of x and t, and that it satisfies the conditions:

$$\lim_{t \to 0} v(x, t) = 0 \quad \text{for} \quad x > 0, \qquad \lim_{x \to 0} v(x, t) = 0 \quad \text{for} \quad t > 0.$$

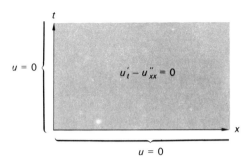

Figure 8.4

8.2.5 Definition
A boundary-value problem is said to be *correctly posed*, if it has the following three properties:
1. There exists a solution.
2. This solution is uniquely determined.
3. The solution is *stable*, i.e. a small change in the boundary values induces only a small change in the solution.

Here property 3 is rather vaguely formulated. In each of the following problems we give a precise meaning to the word stable.

The problem in Example 8.2.4 is not correctly posed, for it does not have property 2. In the next example we study a problem for which property 3 is not satisfied.

8.2.6 Example
Suppose that ε is a "small" positive number. Consider the boundary-value problem:

(a) $$\frac{\partial^2 u}{\partial t^2} + \frac{\partial^2 u}{\partial x^2} = 0, \quad -\infty < x < \infty, \quad t > 0,$$

(b) $$u(x, 0) \quad = 0, \quad -\infty < x < \infty,$$

(c) $$\frac{\partial u(x, 0)}{\partial t} = \varepsilon \sin \frac{x}{\varepsilon}, \quad -\infty < x < \infty.$$

It has the solution $u(x, t) = \varepsilon^2 \sinh (t/\varepsilon) \sin (x/\varepsilon)$ with the property: $\max_{x \in R} u(x, t) = \varepsilon^2 \sinh (t/\varepsilon) \to \infty$ as $\varepsilon \to 0$ for any given $t > 0$. Replace condition (c) by

(c′) $$\frac{\partial u(x, 0)}{\partial t} = 0, \quad -\infty < x < \infty.$$

We get a boundary-value problem that has the solution $u(x, t) = 0$. The second problem then has the property that an arbitrarily small change in the boundary condition (c′), say the change from (c′) to (c), can produce an arbitrarily large change in the solution. We say that the solution of (a), (b), (c′) is not stable. Hence, in practical applications of differential equations, where boundary values are determined by measurements (that cannot be exact), the outcome can be catastrophic if a problem under consideration is not correctly posed.

EXERCISES

805. Suppose that a, b, c are real numbers and that at least one of them is different from zero. Show that the differential equation

$$au''_{xx} + 2bu''_{xy} + cu''_{yy} = 0$$

is

a) elliptic if $ac - b^2 > 0$,
b) parabolic if $ac - b^2 = 0$,
c) hyperbolic if $ac - b^2 < 0$.

806. Determine for each of the following differential equations whether it is elliptic, parabolic, or hyperbolic:

a) $u''_{xx} + u''_{xt} + u''_{tt} + u'_t = e^x$,
b) $u''_{xx} + 4u''_{xt} + 4u''_{tt} - u'_x + u = 0$
c) $u''_{xx} - 4u''_{xt} + u''_{tt} + 2u'_t = \sin x$.

807. Determine for each of the following differential equations at what points it is elliptic, parabolic, hyperbolic:

a) $u''_{xx} + yu''_{xy} + xu''_{yy} = e^{x+y}$,
b) $y^2 u''_{xx} - u''_{yy} + u = 0$,
c) $xu''_{xx} + 2u''_{xy} + yu''_{yy} + xu'_x + u'_y = 0$,
d) $(1 - x^2)u''_{xx} - 2xyu''_{xy} - (1 - y^2)u''_{yy} = xy$.

808. Suppose that f and g are functions in the class $C^2(-\infty, \infty)$. Find a second-order partial differential equation that is satisfied by all functions of the form $u(x, y) = y^3 + f(xy) + g(x)$.

809. Transform the differential equations

a) $2u''_{xx} + 4u''_{xy} + 5u''_{yy} = 0$,
b) $u''_{xx} + 4u''_{xy} + 4u''_{yy} = 0$,
c) $u''_{xx} + 4u''_{xy} - 2u''_{yy} = 0$,

to the form $\lambda_1 u''_{\xi\xi} + \lambda_2 u''_{\eta\eta} = 0$ by convenient rotations of the coordinate axes.

*810. Determine for each of the following differential equations whether it is elliptic, parabolic, or hyperbolic:

a) $u''_{xx} + 2u''_{yy} + u''_{zz} = 2u''_{xy} + 2u''_{yz}$,
b) $u''_{xy}(x, y, z, t) + u''_{zt}(x, y, z, t) = 0$.

811. Find the solution in the first quadrant of the differential equation $(1 + x)u''_{xy} + u'_y = x - 2y$, that satisfies the boundary conditions $u(x, 0) = 1$ and $u(0, y) = \cos y$.

812. a) Prove the assertions for the function $v(x, t)$ of Example 8.2.4.
b) Does the function $v(x, t)$ tend to a limit as the point (x, t) approaches the origin along the curve $x^2 = 4t$?

Hint for (a). By the result of Exercise 802a, the function $t^{-1/2}e^{-x^2/(4t)}$ is a solution of the heat equation for $t > 0$.

813. a) Decide whether the boundary-value problem

$$\frac{\partial u}{\partial t} + \frac{\partial u}{\partial x} = 0, \qquad t > 0, \quad -\infty < x < \infty,$$

$$u(x, 0) = 0, \qquad -\infty < x < \infty,$$

is correctly posed.

b) Do the same for the boundary-value problem

$$t\frac{\partial u}{\partial t} + x\frac{\partial u}{\partial x} = 0, \qquad t > 0, \quad -\infty < x < \infty,$$

$$u(x, 0) = 0, \qquad -\infty < x < \infty.$$

c) Do the same for the boundary-value problem

$$\frac{\partial u}{\partial t} + \frac{\partial u}{\partial x} = u, \qquad t > 0, \quad -\infty < x < \infty,$$

$$u(x, 0) = 0, \qquad -\infty < x < \infty.$$

8.3 THE WAVE EQUATION

8.3.1 Problem

Find the general solution of the wave equation

$$\frac{\partial^2 u}{\partial t^2} = \frac{\partial^2 u}{\partial x^2}. \tag{1}$$

Solution. **1.** Suppose that the function $u(x, t)$ is a solution of (1). It then belongs to the class $C^2(R^2)$, where R^2 denotes the xt-plane. Introduce new independent variables ξ and τ by the equations

$$\xi = x + t, \qquad \tau = x - t,$$

and set

$$v(\xi, \tau) = u\left(\frac{\xi + \tau}{2}, \frac{\xi - \tau}{2}\right).$$

The function $v(\xi, \tau)$ belongs to the class $C^2(R^2)$, where R^2 now denotes the $\xi\tau$-plane. It can be verified (Exercise 814) that

$$\frac{\partial^2 v}{\partial \xi \partial \tau} = 0. \tag{2}$$

Then there is a function $g(\tau)$ in the class $C^1(R)$ and a function $G(\tau)$ in the class $C^2(R)$ such that

$$\frac{\partial v}{\partial \tau} = g(\tau), \qquad \frac{\partial}{\partial \tau}(v - G) = 0.$$

Further there is a function $F(\xi)$ in the class $C^2(R)$ such that

$$v(\xi, \tau) - G(\tau) = F(\xi).$$

It follows that

$$u(x, t) = F(x + t) + G(x - t). \tag{3}$$

2. Suppose that F and G are functions in the class $C^2(R)$. It is then seen that the function $u(x, t)$, defined by (3), satisfies the equation (1).

3. The results in parts 1 and 2 show that the wave equation (1) has the general solution (3), where F and G are arbitrary functions in the class $C^2(R)$.

Consider the families of straight lines $x + t = a$ and $x - t = b$ in the xt-plane, where a and b are parameters. It is seen that the term $F(x + t)$ in the right member of (3) is constant on each particular straight line of the first family; analogously for the term $G(x - t)$.

The next problem has the following physical interpretation. Consider an infinitely long uniform string. Denote by $u(x, t)$ the deviation of the point x of the string from its equilibrium position at the time t. Suppose that the function $u(x, t)$ satisfies equation (1). (There are good physical reasons for this assumption: cf. Exercise 801; the units of length and time are normalized so that the coefficients in both members of (1) are 1.) Suppose that, at the time $t = 0$, the point x has the deviation $u_0(x)$ from its equilibrium position and the velocity $u_1(x)$. Find its deviation at the time $t > 0$.

8.3.2 Problem

Solve the boundary-value problem

$$\frac{\partial^2 u}{\partial t^2} = \frac{\partial^2 u}{\partial x^2}, \qquad x \in R, \quad t > 0, \tag{4a}$$

$$u(x, 0) = u_0(x), \qquad x \in R, \tag{4b}$$

$$\frac{\partial u}{\partial t}(x, 0) = u_1(x), \qquad x \in R, \tag{4c}$$

where $u_0(x)$ and $u_1(x)$ are given functions in the classes $C^2(R)$ and $C^1(R)$ respectively. Show that the problem is correctly posed.

Solution. 1. Suppose that the function $u(x, t)$ is a solution of the problem. By the result in Problem 8.3.1, there are functions f and g in the class $C^2(R)$ such that $u(x, t) = f(x + t) + g(x - t)$. Substitution into (4b) and (4c) gives, setting $c = f(0) - g(0)$,

$$\left.\begin{array}{l} f(x) + g(x) = u_0(x) \\ f'(x) - g'(x) = u_1(x) \end{array}\right\}, \qquad \left.\begin{array}{l} f(x) + g(x) = u_0(x) \\ f(x) - g(x) = \displaystyle\int_0^x u_1(\xi)\, d\xi + c \end{array}\right\},$$

$$\left.\begin{array}{l} 2f(x) = u_0(x) + \displaystyle\int_0^x u_1(\xi)\, d\xi + c \\[2mm] 2g(x) = u_0(x) - \displaystyle\int_0^x u_1(\xi)\, d\xi - c \end{array}\right\},$$

$$u(x, t) = \tfrac{1}{2}[u_0(x + t) + u_0(x - t)] + \tfrac{1}{2} \int_{x-t}^{x+t} u_1(\xi)\, d\xi. \qquad (5)$$

2. It can be verified (Exercise 815) that the function (5) satisfies the boundary-value problem (4).

3. The results in parts **1** and **2** show that the problem (4) has the unique solution (5). Suppose that ε is a positive number and $\varepsilon_0(x)$ and $\varepsilon_1(x)$ are functions in the classes $C^2(R)$ and $C^1(R)$ respectively such that $|\varepsilon_0(x)| < \varepsilon/2$ and $\left|\int_0^x \varepsilon_1(\xi)\, d\xi\right| < \varepsilon/2$ for all $x \in R$. Add these functions to the right members of (4b) and (4c) respectively. We get a new boundary-value problem. Let $v(x, t)$ be the solution of this new problem. By formula (5) we have

$$|v(x, t) - u(x, t)| < \varepsilon \qquad \text{for} \quad x \in R \quad \text{and} \quad t > 0.$$

We then say that the solution is stable. Hence problem (4) is correctly posed.

8.3.3 Definition

Formula (5) is called the *d'Alembert formula*. The formula shows that the value of the solution $u(x, t)$ at the point (x, t), $t > 0$, depends only on the values of the functions u_0 and u_1 on the interval $[x - t, x + t]$.

The next problem has the following physical interpretation. Suppose (Figure 8.5) that a uniform string has its end points at the points 0 and π of the x-axis, that its vibrations satisfy the wave equation (1), that position and velocity are given for each point x of the string at the time $t = 0$, and that the string can turn without friction around its end points

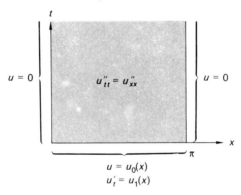

Figure 8.5

(this last assumption gives the condition below on the second derivative at the points 0 and π). Find the motion of the string for $t > 0$.

We shall solve this problem by a method that is called *separation of variables*. There is a shorter solution (see Exercise 816). The method of separation of variables, however, has a wider applicability, and this problem seems convenient for a presentation of the method.

We postpone proving that Problem 8.3.4 is correctly posed (see Problem 8.3.6).

8.3.4 Problem

Find a solution of the boundary-value problem:

$$\frac{\partial^2 u}{\partial t^2} = \frac{\partial^2 u}{\partial x^2}, \qquad 0 < x < \pi, \quad t > 0, \tag{6a}$$

$$u(x, 0) = u_0(x), \qquad 0 < x < \pi, \tag{6b}$$

$$\frac{\partial u}{\partial t}(x, 0) = u_1(x), \qquad 0 < x < \pi, \tag{6c}$$

$$u(0, t) = u(\pi, t) = 0, \qquad t > 0, \tag{6d}$$

where $u_0(x)$ and $u_1(x)$ are given functions in the classes $C^2[0, \pi]$ and $C^1[0, \pi]$ respectively, such that $u_0(0) = u_0(\pi) = 0, u_1(0) = u_1(\pi) = 0$, and $u_0''(0) = u_0''(\pi) = 0$.

Solution. **1.** Suppose that there exists a function $X(x)$, $0 \leqslant x \leqslant \pi$, with only a finite number of zeros (the points $x_0, \ldots, x_m$ in Fig. 8.6), and a function $T(t)$, $t \geqslant 0$, with only isolated zeros (the points $t_0, \ldots, t_n \ldots$ in Fig. 8.6), such that the function

$$v(x, t) = X(x)T(t)$$

satisfies conditions (6a) and (6d). Then

(a) $$X(0) = X(\pi) = 0.$$

Let Ω_k be a rectangular region, bounded by straight lines $x = x_\mu$, $x = x_{\mu+1}$, $t = t_\nu$, $t = t_{\nu+1}$ (Fig. 8.6). Then $v(x, t) \neq 0$ in Ω_k. There is a constant λ such that, in Ω_k,

$$X(x)T''(t) = X''(x)T(t),$$

$$\frac{X''(x)}{X(x)} = \frac{T''(t)}{T(t)} = -\lambda,$$

(b, c) $$X'' + \lambda X = 0, \qquad T'' + \lambda T = 0.$$

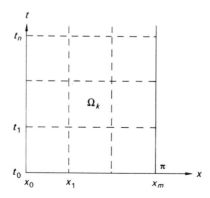

Figure 8.6

To any two adjoining regions Ω_k, separated by a vertical segment in Fig. 8.6, there corresponds the same value of λ, for $T''(t)/T(t)$ takes the same constant value in the two regions. Analogously the same value of λ corresponds to any two adjoining regions, separated by a horizontal segment. Therefore equations (b) and (c) hold throughout the region $0 < x < \pi, t > 0$. Equations (a) and (b) and the result in Section 2.6.1 show that there exists a positive integer n such that $\lambda = n^2$ and $X(x) = \sin nx$ (times a nonzero constant that we let be absorbed into the constants a_n and b_n below). Equation (c) then shows that the function $T(t)$ is of the form $a_n \cos nt + b_n \sin nt$ where a_n and b_n are constants, not both zero. Hence

(d) $$v(x, t) = (a_n \cos nt + b_n \sin nt) \sin nx.$$

It is readily verified that the function (d) satisfies the conditions (6a) and (6d).

2. It is also readily verified that every function $w(x, t)$ of the form

$$w(x, t) = \sum_{v=1}^{n} (a_v \cos vt + b_v \sin vt) \sin vx$$

satisfies the conditions (6a) and (6d).

3. Trying to satisfy also conditions (6b) and (6c), we introduce the series

$$\sum_{n=1}^{\infty} (a_n \cos nt + b_n \sin nt) \sin nx,$$

where the coefficients a_n and b_n are determined by expanding the functions $u_0(x)$ and $u_1(x)$ in sine series (see Example 2.2.6):

$$u_0(x) \sim \sum_{n=1}^{\infty} a_n \sin nx \quad \text{and} \quad u_1(x) \sim \sum_{n=1}^{\infty} nb_n \sin nx.$$

Let $\tilde{u}_0(x)$ denote the odd function with domain R and with period 2π that coincides with $u_0(x)$ for $0 \leqslant x \leqslant \pi$. Define $\tilde{u}_1(x)$ analogously. Theorem 1.4.4 then shows that, for all $x \in R$,

(e) $$\tilde{u}_0(x) = \sum_{n=1}^{\infty} a_n \sin nx \quad \text{and} \quad \tilde{u}_1(x) = \sum_{n=1}^{\infty} nb_n \sin nx.$$

The two functions (e) belong to the classes $C^2(R)$ and $C^1(R)$ respectively. Problem 8.3.2 with the functions (e) as boundary functions has the following solution (the d'Alembert formula (5) is applied; the term-by-term integration is legitimate by Theorem 2.4.3):

$$\tilde{u}(x, t) = \tfrac{1}{2}[\tilde{u}_0(x + t) + \tilde{u}_0(x - t)] + \tfrac{1}{2} \int_{x-t}^{x+t} \tilde{u}_1(\xi) \, d\xi$$

$$= \sum_{n=1}^{\infty} a_n \frac{\sin n(x + t) + \sin n(x - t)}{2} - \sum_{n=1}^{\infty} b_n \frac{\cos n(x + t) - \cos n(x - t)}{2}$$

$$= \sum_{n=1}^{\infty} a_n \sin nx \cos nt + \sum_{n=1}^{\infty} b_n \sin nx \sin nt.$$

It is now seen that the function

$$u(x, t) = \sum_{n=1}^{\infty} (a_n \cos nt + b_n \sin nt) \sin nx \tag{7}$$

is a solution of the problem (6).

In order to show that Problem 8.3.4 is correctly posed we introduce the notion of energy and prove a property of this notion.

8.3.5 Remark

The *energy* $E(t)$ of the solution (7) for the problem (6) at the time $t \geqslant 0$ is defined by the integral

$$E(t) = \int_0^\pi \left[\left(\frac{\partial u}{\partial t} \right)^2 + \left(\frac{\partial u}{\partial x} \right)^2 \right] dx. \tag{8}$$

(The first term of the integrand is the square of the velocity of a vibrating string at the point x, and the second term represents the dilatation of the string at the same point; $E(t)$ then is a sum of kinetic energy and potential energy.)   We claim that $E(t)$ is constant.   This follows from (observe that $\partial u/\partial t = 0$ for $x = 0$ and $x = \pi$; formula (6a) gives the second equality sign):

$$\frac{dE}{dt} = \int_0^\pi 2 \left(\frac{\partial u}{\partial t} \frac{\partial^2 u}{\partial t^2} + \frac{\partial u}{\partial x} \frac{\partial^2 u}{\partial x \partial t} \right) dx$$

$$= 2 \int_0^\pi \left(\frac{\partial u}{\partial t} \frac{\partial^2 u}{\partial x^2} + \frac{\partial u}{\partial x} \frac{\partial^2 u}{\partial x \partial t} \right) dx$$

$$= 2 \int_0^\pi \frac{\partial}{\partial x} \left(\frac{\partial u}{\partial t} \frac{\partial u}{\partial x} \right) dx = 2 \left[\frac{\partial u}{\partial t} \frac{\partial u}{\partial x} \right]_{x=0}^{x=\pi} = 0.$$

8.3.6 Problem

Show that the boundary-value problem (6) is correctly posed.

Solution. 1.   Suppose that the functions $v(x, t)$ and $w(x, t)$ are solutions of the boundary-value problem (6).   Then the function $u(x, t) = v(x, t) - w(x, t)$ satisfies the conditions (6a) and (6d) and, for $0 < x < \pi$, the equations $u(x, 0) = 0$ and $\partial u(x, 0)/\partial t = 0$.   Let $E(t)$ denote the energy of $u(x, t)$.   Then $E(0) = 0$, and the remark above gives for $t > 0$ and $0 < x < \pi$:

$$E(t) = 0, \qquad \left(\frac{\partial u}{\partial t} \right)^2 + \left(\frac{\partial u}{\partial x} \right)^2 = 0,$$

$$u(x, t) = u(0, t) + \int_0^x \frac{\partial u}{\partial x} (\xi, t) \, d\xi = 0,$$

$$v(x, t) - w(x, t) = 0.$$

Hence the boundary-value problem (6) has the function (7) as its only solution.

2. Suppose that $\varepsilon > 0$. Suppose that the functions $\varepsilon_0(x)$ and $\varepsilon_1(x)$ fulfill the same conditions as the function $u_0(x)$ and $u_1(x)$ respectively in Problem 8.3.4, and that $\int_0^\pi [\varepsilon_0'^2(x) + \varepsilon_1^2(x)]\, dx < \varepsilon^2/\pi$. Add $\varepsilon_0(x)$ and $\varepsilon_1(x)$ to the right members of (6b) and (6c) respectively. Then a new boundary-value problem is obtained. Let $v(x, t)$ be the solution of this problem. Set $\eta(x, t) = v(x, t) - u(x, t)$, and denote the energy of $\eta(x, t)$ by $E(t)$. By the Schwarz inequality (Lemma 2.4.1) and the remark above we have for $0 < x < \pi$ and $t > 0$:

$$\left| v(x, t) - u(x, t) \right| = \left| \eta(x, t) \right| = \left| \eta(0, t) + \int_0^x \frac{\partial \eta}{\partial x}(\xi, t)\, d\xi \right|$$

$$\leqslant \int_0^\pi \left| \frac{\partial \eta}{\partial x}(x, t) \right| dx \leqslant \left[\pi \int_0^\pi \left(\frac{\partial \eta}{\partial x}(x, t) \right)^2 dx \right]^{1/2}$$

$$\leqslant [\pi E(t)]^{1/2} = [\pi E(0)]^{1/2} < \varepsilon.$$

We then say that the solution of problem (6) is stable.
The solution of Problem 8.3.6 is now complete.

EXERCISES

Exercises 818–821 are not quite correct. The boundary conditions do not satisfy all the conditions in Problem 8.3.4. This has the effect that the "solutions" obtained do not satisfy the wave equation on certain straight lines in the xt-plane. The exercises have been inserted to give the reader practice in the method of separation of variables in exercises with fairly short computations. A reader who dislikes such defective exercises can, e.g., replace the function $u(x, 0) = \pi x - x^2$ in the first boundary condition of Exercise 818 by a trigonometric polynomial $t(x) = \sum_{v=1}^n b_v \sin vx$ such that $|\pi x - x^2 - t(x)| < \varepsilon$ for $0 < x < \pi$, where ε is a given positive number. Then he has a correct exercise. Similarly for Exercises 819–821.

814. Prove equation (2).

815. Show that the function (5) satisfies the boundary-value problem (4).

816. Solve Problem 8.3.4 by substituting into the d'Alembert formula (5) the extensions of the functions $u_0(x)$ and $u_1(x)$ that are odd functions and have period 2π.

817. Suppose that $c > 0$. Find the solution of the wave equation $u_{tt}'' = c^2 u_{xx}''$, $x \in R$, $t > 0$, that has the boundary values $u(x, 0) = (1 + x^2)^{-1}$ and $u_t'(x, 0) = \sin x$ for all $x \in R$.

818. Find a solution of the boundary-value problem:

$$\begin{aligned}
u_{tt}'' &= u_{xx}'', & 0 < x < \pi, \quad t > 0, \\
u(x, 0) &= \pi x - x^2, & 0 < x < \pi, \\
u_t'(x, 0) &= 0, & 0 < x < \pi, \\
u(0, t) &= u(\pi, t) = 0, & t > 0.
\end{aligned}$$

819. Change the second condition in the previous exercise to $u(x, 0) = 2\pi x - x^2$ and the last condition to $u(0, t) = u_x'(\pi, t) = 0$. Find a solution $u(x, t)$.

820. Find $u(\frac{3}{4}, \frac{3}{2})$, where $u(x, t)$ is the solution of the wave equation $u''_{tt} = u''_{xx}$, $0 < x < 1$, $t > 0$, with the boundary conditions $u(0, t) = u(1, t) = 0$, $t > 0$, and
   a) $u(x, 0) = 0$ and $u'_t(x, 0) = x(1 - x)$, $0 < x < 1$,
   b) $u(x, 0) = x(1 - x)$ and $u'_t(x, 0) = x^2(1 - x)$, $0 < x < 1$.

821. Find a solution of the boundary-value problem

$$u''_{tt} = u''_{xx}, \qquad\qquad 0 < x < 1, \quad t > 0,$$
$$u(x, 0) = 0, \qquad\qquad 0 < x < 1,$$
$$u'_t(x, 0) = 2x - 1, \qquad\qquad 0 < x < 1,$$
$$u'_x(0, t) = u'_x(1, t) = 0, \qquad t > 0.$$

822. Let $E_k(t)$, $E_p(t)$ and $E(t)$ denote the kinetic energy, the potential energy and the total energy respectively of the solution of the boundary-value problem

$$u''_{tt} = u''_{xx}, \qquad\qquad 0 < x < \pi, \quad t > 0,$$
$$u(x, 0) = \sin x, \qquad\qquad 0 < x < \pi,$$
$$u'_t(x, 0) = 0, \qquad\qquad 0 < x < \pi,$$
$$u(0, t) = u(\pi, t) = 0, \qquad t > 0.$$

Sketch the three curves $y = E_k(t)$, $y = E_p(t)$, $y = E(t)$ for $0 \leqslant t \leqslant 2\pi$.

*823. Consider the differential equation $u''_{xx} - 2u''_{xt} + u''_{tt} = 0$.
   a) Is it elliptic, parabolic, or hyperbolic?
   b) Find its general solution.
   Consider the boundary-value problem

$$u''_{xx} - 2u''_{xt} + u''_{tt} = 0, \qquad x \geqslant 0, \quad t \geqslant 0,$$
$$u(x, 0) = 2x^2, \qquad x \geqslant 0,$$
$$u(0, t) = 0, \qquad t \geqslant 0.$$

   c) Solve this problem.
   d) Is it correctly posed?
   Hint for (b). Use the same substitution as in the solution of Problem 8.3.1.

824. Suppose that the functions $u_0(x)$ and $u_1(x)$ are given functions in the classes $C^2(R)$ and $C^1(R)$ respectively, and that the functions $f(x, t)$ and $f'_x(x, t)$ are continuous for $x \in R$, $t > 0$. Show that the boundary-value problem

$$u''_{tt} - u''_{xx} = f(x, t), \qquad x \in R, \quad t > 0,$$
$$u(x, 0) = u_0(x) \quad \text{and} \quad u'_t(x, 0) = u_1(x), \qquad x \in R,$$

has exactly one solution, and that this solution is

$$u(x, t) = \tfrac{1}{2}[u_0(x + t) + u_0(x - t)] + \tfrac{1}{2} \int_{x-t}^{x+t} u_1(\xi)\, d\xi$$

$$+ \tfrac{1}{2} \int_0^t \int_{x-(t-\tau)}^{x+(t-\tau)} f(\xi, \tau)\, d\xi\, d\tau.$$

.

825. Find $u(\tfrac{1}{4}, 1)$, where $u(x, t)$ is the solution of the boundary-value problem

$$u''_{tt} - u''_{xx} = t \sin^2 \pi x, \qquad 0 < x < 1, \quad t > 0,$$
$$u(x, 0) = u'_t(x, 0) = 0, \qquad 0 < x < 1,$$
$$u(0, t) = u(1, t) = 0, \qquad t > 0.$$

826. Suppose that $c > 0$. Consider the differential equation $c^2(u''_{xx} + u''_{yy}) = u''_{tt}$. Find ordinary differential equations for the functions $X(x)$, $Y(y)$ and $T(t)$, when the function $u(x, y, t) = X(x)Y(y)T(t)$ is a solution.

827. Suppose that the two functions (e) belong to the classes $C^4(R)$ and $C^3(R)$ respectively. Show by substitution that the function (7) satisfies equation (6a).
    Hint. Apply the theorems of Section 1.2.4 and a variant of Lemma 1.5.1.

8.4 THE HEAT EQUATION

In this section (8.4) we study four boundary-value problems for the heat equation. The first three problems (8.4.1, 8.4.4, 8.4.5) have the physical interpretation that the functions $u_0(x)$ and $u(x, t)$ give the temperature at the point x of a thin uniform rod (of finite or infinite length) at the time $t = 0$ and at any time $t > 0$ respectively. The rod has no exchange of heat with its surroundings except at its end point or end points (if there are any). In the three problems the number of end points is zero, two and one respectively. In the second problem, both end points of the rod are kept at the temperature zero for $t > 0$. In the third problem, the temperature at the end point is a given function of time. In the problems it is required to find solutions that are continuous and bounded on a closed domain (Example 8.2.4 shows that the first of these conditions is necessary; a result in [3], pp. 30–31, shows that the second of these conditions (or some similar condition) is necessary). We shall see that Fourier and Laplace transforms sometimes can be used to reduce the solution of a boundary-value problem for a partial differential equation to a problem for an ordinary differential equation. We postpone showing that Problem 8.4.1 is correctly posed till we have studied an auxiliary theorem (Theorem 8.4.2, called the maximum principle for the heat equation). We divide the solution of each problem into several steps. In the first step we introduce some extra assumptions in order to obtain a function $u(x, t)$ that may be a solution. In the second step we reject these extra assumptions and prove that, in fact, $u(x, t)$ is a solution.

8.4.1 Problem

Find a solution of the boundary-value problem

$$\frac{\partial u}{\partial t} = \frac{\partial^2 u}{\partial x^2}, \qquad x \in R, \quad t > 0, \tag{1a}$$

$$u(x, 0) = u_0(x), \qquad x \in R, \tag{1b}$$

$$u(x, t) \text{ is continuous for } x \in R, \qquad t \geqslant 0, \tag{1c}$$

$$u(x, t) \text{ is bounded for } x \in R, \qquad t \geqslant 0, \tag{1d}$$

where $u_0(x)$ is a given function, continuous and bounded for $x \in R$.

Solution. **1.** Suppose that the function $u(x, t)$ is a solution, that $u(x, t)$ has a Fourier transform $U(\xi, t)$ for each $t > 0$, and that differentiation with respect to t under the integral sign in the expression for the Fourier transform is legitimate:

$$U(\xi, t) = \frac{1}{2\pi} \int_{-\infty}^{\infty} u(x, t)\, e^{-i\xi x}\, dx,$$

$$\frac{\partial U(\xi, t)}{\partial t} = \frac{1}{2\pi} \int_{-\infty}^{\infty} \frac{\partial u(x, t)}{\partial t}\, e^{-i\xi x}\, dx = \frac{1}{2\pi} \int_{-\infty}^{\infty} \frac{\partial^2 u(x, t)}{\partial x^2}\, e^{-i\xi x}\, dx.$$

For the last integral, suppose that the formula

$$\int_{-\infty}^{\infty} f'g\, dx = \left[f g \right]_{-\infty}^{\infty} - \int_{-\infty}^{\infty} f g'\, dx = - \int_{-\infty}^{\infty} f g'\, dx$$

is applicable twice:

$$\frac{\partial U(\xi, t)}{\partial t} = (i\xi)\frac{1}{2\pi} \int_{-\infty}^{\infty} \frac{\partial u(x, t)}{\partial x}\, e^{-i\xi x}\, dx$$

$$= (i\xi)^2 \frac{1}{2\pi} \int_{-\infty}^{\infty} u(x, t)e^{-i\xi x}\, dx = -\xi^2 U(\xi, t).$$

Suppose that $u_0(x)$ has a Fourier transform $U_0(\xi)$, and that $U(\xi, t) \to U_0(\xi)$ as $t \to 0$ for each $\xi \in R$. Then, given a real number ξ, we have an initial-value problem for an ordinary differential equation:

$$\frac{\partial U(\xi, t)}{\partial t} + \xi^2 U(\xi, t) = 0,$$

$$U(\xi, 0) = \frac{1}{2\pi} \int_{-\infty}^{\infty} u_0(x)e^{-i\xi x}\, dx.$$

This problem has the solution

$$U(\xi, t) = U(\xi, 0)e^{-\xi^2 t}.$$

The result in Example 4.4.12 shows that

$$\mathscr{F}[e^{-x^2/2}]\,(\xi) = \frac{1}{\sqrt{2\pi}}\, e^{-\xi^2/2}.$$

Suppose $a > 0$. The result in Exercise 417c then gives

$$\mathscr{F}[e^{-a^2 x^2/2}]\,(\xi) = \frac{1}{a\sqrt{2\pi}}\, e^{-\xi^2/(2a^2)}.$$

Letting $a^2 = 1/(2t)$ we obtain:

$$\mathscr{F}[e^{-x^2/(4t)}]\,(\xi) = \sqrt{\frac{t}{\pi}} \cdot e^{-\xi^2 t}.$$

Suppose that the convolution theorem and the inversion formula for Fourier transforms (Theorem 4.5.5 and formula (3) of Section 4.4.2) are applicable. Then $u(x, t)$ is equal to $1/(2\pi)$ times the convolution of the functions $u_0(x)$ and $(\pi/t)^{1/2}e^{-x^2/(4t)}$:

$$u(x, t) = u_0(x) * \frac{1}{\sqrt{4\pi t}}\, e^{-x^2/(4t)},$$

$$u(x, t) = \frac{1}{\sqrt{4\pi t}} \int_{-\infty}^{\infty} u_0(y)\, e^{-(x-y)^2/(4t)}\, dy. \tag{2}$$

2. Now suppose only that the function $u_0(x)$ is continuous and bounded for $x \in R$. (Omit all the extra assumptions in part **1**.) Define the function $u(x, t)$ for $x \in R$ and $t > 0$ by (2). It can be verified that $u(x, t)$ satisfies equation (1a) (Exercise 828; use of the technique in the proof of Theorem 5.1.4 shows that differentiations under the integral sign are legitimate). The substitution $y = x + 2\sqrt{tz}$ in (2) gives

$$u(x, t) = \frac{1}{\sqrt{\pi}} \int_{-\infty}^{\infty} u_0(x + 2\sqrt{tz})e^{-z^2}\, dz. \tag{3}$$

Suppose that ε and M are positive numbers, that $|u_0(x)| < M$ for all $x \in R$, and that ω is a positive number such that

$$\frac{1}{\sqrt{\pi}} \int_{|z|>\omega} 2Me^{-z^2}\, dz < \frac{\varepsilon}{2}.$$

Suppose that $x_1 \in R$. We have, using formula (9) of Example 4.4.12,

$$u(x, t) - u_0(x_1) = \frac{1}{\sqrt{\pi}} \int_{|z|<\omega} + \int_{|z|>\omega} [u_0(x + 2\sqrt{tz})$$
$$- u_0(x_1)]e^{-z^2}\, dz = I_1 + I_2,$$

where I_1 and I_2 denote the integrals on $|z| < \omega$ and $|z| > \omega$ respectively. Now $|I_2| < \varepsilon/2$. There is a positive number δ such that, for $|x - x_1| < \delta, 0 < t < \delta$, $|z| < \omega$,

$$\left|u_0(x + 2\sqrt{tz}) - u_0(x_1)\right| < \frac{\varepsilon}{2}, \qquad \text{and hence } |I_1| < \frac{\varepsilon}{2}.$$

Then, for $|x - x_1| < \delta$ and $0 < t < \delta$,

$$\left|u(x, t) - u_0(x_1)\right| < \varepsilon.$$

It follows that the function $u(x, t)$ also fulfills conditions (1b) and (1c). Finally condition (1d) is fulfilled, for (3) and formula (9) of Example 4.4.12 show that $|u(x, t)| < M$ for $x \in R$, $t \geqslant 0$. Hence $u(x, t)$ is a solution of problem (1).

The following theorem is a mathematical counterpart of the physical fact that heat flows from a warmer place to a colder place.

8.4.2 Theorem (the maximum principle for the heat equation). *Suppose that Ω is the region $a < x < b$, $0 < t < T$, in the xt-plane, that $\partial\Omega$ is the boundary of Ω, that Ω^- is the union of Ω and $\partial\Omega$, that S is the part of $\partial\Omega$ situated on the lines $t = 0$, $x = a$ and $x = b$, and that S' is the rest of $\partial\Omega$ (i.e. the set $a < x < b$, $t = T$; see* Figure 8.7). *Suppose that $u(x, t)$ is a continuous function with domain Ω^-, and that $u(x, t)$ satisfies the heat equation $u'_t = u''_{xx}$ on the set $\Omega \cup S'$. Then $u(x, t)$ takes on its largest value, and also its smallest value, on S.*

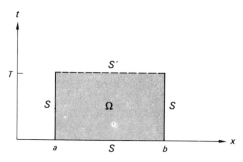

Figure 8.7

Proof. Suppose $\varepsilon > 0$. Set

$$v(x, t) = u(x, t) + \varepsilon x^2.$$

Then

(a)
$$\frac{\partial v}{\partial t} - \frac{\partial^2 v}{\partial x^2} = -2\varepsilon \qquad \text{for} \quad (x, t) \in \Omega \cup S'.$$

Suppose that $(x_0, t_0) \in \Omega^-$ and that

$$v(x_0, t_0) = \max_{\Omega^-} v(x, t).$$

It is seen that

$$(x_0, t_0) \in \Omega \Rightarrow \frac{\partial v}{\partial t}(x_0, t_0) = 0 \qquad \text{and} \qquad \frac{\partial^2 v}{\partial x^2}(x_0, t_0) \leqslant 0,$$

$$(x_0, t_0) \in S' \Rightarrow \frac{\partial v}{\partial t}(x_0, t_0) \geqslant 0 \qquad \text{and} \qquad \frac{\partial^2 v}{\partial x^2}(x_0, t_0) \leqslant 0.$$

In each case we get a contradiction to (a). Hence $(x_0, t_0) \in S$. Set $M = \max\limits_{S} u(x, t)$. Then

$$\max_{\Omega^-} u(x, t) \leqslant \max_{\Omega^-} v(x, t) = \max_{S} v(x, t) \leqslant M + \varepsilon \max(a^2, b^2),$$

$$\max_{\Omega^-} u(x, t) \leqslant M, \qquad \max_{\Omega^-} u(x, t) = M,$$

which is the first assertion. The second assertion is obtained by applying the first
assertion to the function $-u(x, t)$.

8.4.3 Problem
Show that the boundary-value problem (1) is correctly posed.

Solution. **1.** We already know that the problem has the solution (2). Suppose that
the problem has two solutions. Let $u(x, t)$ be their difference. Then $u(x, t)$ satisfies
the conditions (1) with the exception that (1b) is replaced by $u(x, 0) = 0$, $x \in R$.
Suppose that M is a positive number, such that $|u(x, t)| < M$ for $x \in R$ and $t > 0$.
Suppose that $a > 0$. Then the values of the function $u(x, t) - Mx^2/a^2 - 2Mt/a^2$
are negative or zero on the boundary of the region $-a < x < a, t > 0$. Further
this function is a solution of the heat equation in the half-plane $t > 0$. The maximum
principle then shows that

$$u(x, t) - Mx^2/a^2 - 2Mt/a^2 \leqslant 0 \qquad \text{for} \quad -a < x < a, \quad t > 0.$$

Fix the point (x, t), and let a tend to infinity. It follows that $u(x, t) \leqslant 0$. Hence the
function $u(x, t)$ does not take on positive values. Application of this result to the
function $-u(x, t)$ shows that $u(x, t)$ does not take on negative values. It follows
that the function (2) is the only solution of the problem.

2. Let $u(x, t)$ be the function (2). Suppose that $\varepsilon > 0$, that the function $\varepsilon_0(x)$
is continuous and bounded for $x \in R$, and that $|\varepsilon_0(x)| < \varepsilon$ for $x \in R$. Add $\varepsilon_0(x)$
to the right member of (1b). Then a new boundary-value problem is obtained.
Let $v(x, t)$ be its solution. Then formula (3) above and formula (9) of Example
4.4.12 show that

$$|u(x, t) - v(x, t)| < \varepsilon \qquad \text{for} \quad x \in R, \quad t > 0.$$

We then say that the solution (2) is stable.
Summing up, problem (1) is correctly posed.

8.4.4 Problem
Solve the boundary-value problem

$$\frac{\partial u}{\partial t} = \frac{\partial^2 u}{\partial x^2}, \qquad 0 < x < \pi, \quad t > 0, \tag{4a}$$

$$u(x, 0) = u_0(x), \qquad 0 < x < \pi, \tag{4b}$$

$$u(0, t) = u(\pi, t) = 0, \qquad t > 0, \tag{4c}$$

$$u(x, t) \quad \text{is continuous for} \quad 0 \leqslant x \leqslant \pi, \qquad t \geqslant 0, \tag{4d}$$

where the function $u_0(x)$ belongs to the class $C^2[0, \pi]$ and has the properties:
$u_0(0) = u_0(\pi) = u_0''(0) = u_0''(\pi) = 0$. Show that the problem is correctly posed.

Solution. **1.** We argue as in part **1** of the solution of Problem 8.3.4. We write down
the formulas and leave it to the reader to fill in accompanying comments:

$$v(x, t) = X(x)T(t), \qquad X(0) = X(\pi) = 0,$$

$$X(x)T'(t) = X''(x)T(t), \qquad \frac{X''(x)}{X(x)} = \frac{T'(t)}{T(t)} = -\lambda,$$

$$X'' + \lambda X = 0, \qquad T' + \lambda T = 0,$$

$$\lambda = n^2, \quad n \in Z^+, \qquad X(x) = \sin nx, \qquad T(t) = a_n e^{-n^2 t},$$

$$v(x, t) = a_n e^{-n^2 t} \sin nx,$$

(b)
$$u_0(x) \sim \sum_{n=1}^{\infty} a_n \sin nx,$$

$$u(x, t) = \sum_{n=1}^{\infty} a_n e^{-n^2 t} \sin nx. \tag{5}$$

2. The sum of the series in (b) is a function in the class $C^2(R)$. Then, using Lemma 1.5.1, it is seen that the series $\sum_{n=1}^{\infty} |a_n|$ is convergent. Variants for the half-plane $x \in R$, $t \geq 0$, of theorems in Section 1.2.4 now show that the function $u(x, t)$, defined by (5), fulfills condition (4d). It is seen that $u(x, t)$ also satisfies the remaining conditions (4). Hence $u(x, t)$ is a solution of problem (4).

3. Suppose that problem (4) has two solutions. Let $u(x, t)$ be their difference. Then $u(x, t)$ satisfies conditions (4) with the exception that (4b) is replaced by $u(x, 0) = 0$, $0 < x < \pi$. The maximum principle now shows that $u(x, t) = 0$ for $0 \leq x \leq \pi, t \geq 0$. It follows that the function (5) is the only solution of the problem.

4. Let $u(x, t)$ be the function (5). Suppose that $\varepsilon > 0$, that the function $\varepsilon_0(x)$ satisfies the same conditions as the function $u_0(x)$, and that $|\varepsilon_0(x)| < \varepsilon$ for $0 < x < \pi$. Add $\varepsilon_0(x)$ to the right member of (4b). Then a new boundary-value problem is obtained. Let $v(x, t)$ be its solution. The maximum principle shows that

$$|u(x, t) - v(x, t)| < \varepsilon \qquad \text{for} \quad 0 < x < \pi, \quad t > 0.$$

We then say that the solution (5) is stable.

Summing up, problem (4) has the solution (5), and the problem is correctly posed.

8.4.5 Problem
Solve the boundary-value problem (Fig. 8.8)

$$\frac{\partial u}{\partial t} = \frac{\partial^2 u}{\partial x^2}, \qquad x > 0, \quad t > 0, \tag{6a}$$

$$u(x, 0) = 0, \qquad x > 0, \tag{6b}$$

$$u(0, t) = f(t), \qquad t > 0, \tag{6c}$$

$$u(x, t) \quad \text{is continuous for} \quad x \geq 0, \qquad t \geq 0, \tag{6d}$$

$$u(x, t) \quad \text{is bounded for} \quad x \geq 0, \qquad t \geq 0, \tag{6e}$$

where $f(t)$ is a bounded function in the class $C^1[0, \infty)$, $f(0) = 0$, and $f'(t)$ is a bounded function. Show that the problem is correctly posed.

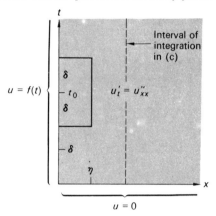

Figure 8.8

Solution. **1.** Suppose that the function $u(x, t)$ is a solution of the problem. Given $x > 0$, denote the Laplace transform of $u(x, t)$ by $U(x, s)$ and the Laplace transform of $f(t)$ by $F(s)$:

(c)
$$\mathcal{L}[u(x, t)](s) = U(x, s) = \int_0^\infty e^{-st} u(x, t)\, dt,$$

$$\mathcal{L}[f(t)](s) = F(s) = \int_0^\infty e^{-st} f(t)\, dt,$$

where we let s be a real positive variable. Suppose that, given $s > 0$, it is legitimate to differentiate twice with respect to x under the integral sign in (c):

$$\frac{\partial^2 U(x, s)}{\partial x^2} = \int_0^\infty e^{-st} \frac{\partial^2 u(x, t)}{\partial x^2}\, dt = \int_0^\infty e^{-st} \frac{\partial u(x, t)}{\partial t}\, dt$$

$$= \left[e^{-st} u(x, t) \right]_{t=0}^{t=\infty} + s \int_0^\infty e^{-st} u(x, t)\, dt = sU(x, s).$$

The assumptions imply that $U(x, s) \to F(s)$ as $x \to 0$. We then have an initial-value problem for an ordinary differential equation:

$$\frac{\partial^2 U(x, s)}{\partial x^2} = sU(x, s),$$

$$U(0, s) = F(s).$$

The differential equation has the general solution

$$U(x, s) = c_1(s)e^{x\sqrt{s}} + c_2(s)e^{-x\sqrt{s}},$$

where the coefficients depend on s. Suppose that $c_1(s) = 0$ for all $s > 0$ (this can always be achieved by appropriately adjusting $c_2(s)$; further this assumption is convenient, for $e^{-x\sqrt{s}}$ is a Laplace transform by Exercise 519 while $e^{x\sqrt{s}}$ is not a Laplace transform because it does not tend to zero as $s \to \infty$). We then have

$$U(x, s) = F(s)e^{-x\sqrt{s}}.$$

The convolution theorem for Laplace transforms (Theorem 5.3.5) and the result in Exercise 519 now show that

$$u(x, t) = \int_0^t f(t - y) \frac{x}{2\sqrt{\pi y^3}} e^{-x^2/(4y)} \, dy. \tag{7}$$

2. Define the function $u(x, t)$ for $x > 0$, $t > 0$, by (6). The substitution $y = x^2/(4z^2)$ gives

$$u(x, t) = \frac{2}{\sqrt{\pi}} \int_{x/\sqrt{4t}}^{\infty} f(t - x^2/(4z^2)) \, e^{-z^2} \, dz. \tag{8}$$

Further define $u(x, t)$ on the boundary of the first quadrant by continuity (we shall see that this definition makes sense). It is left to the reader, using the expressions (7) and (8) for positive values of x and t, to verify properties (6a), (6b), (6e) and the part of (6d) which states that $u(x, t)$ is continuous for $x > 0$, $t \geqslant 0$, and for $x = t = 0$ (Exercise 831). It remains to verify property (6c) and the part of (6d) which states that $u(x, t)$ is continuous for $x = 0$, $t > 0$. To this end, let t_0 and ε denote given positive numbers. Suppose that M and δ are positive numbers such that

$$2\delta < t_0, \qquad |f(t)| < M \qquad \text{for} \quad t \geqslant 0,$$

$$|f(t) - f(t_0)| < \frac{\varepsilon}{3} \qquad \text{for} \quad |t - t_0| < 2\delta.$$

We have for $x > 0$, $t > 0$,

$$u(x, t) - f(t_0) = \int_0^t f(t - y) \frac{x}{2\sqrt{\pi y^3}} e^{-x^2/(4y)} \, dy - \int_0^\infty f(t_0) \frac{x}{2\sqrt{\pi y^3}} e^{-x^2/(4y)} \, dy$$

$$= \int_0^\delta + \int_\delta^t [f(t - y) - f(t_0)] \frac{x}{2\sqrt{\pi y^3}} e^{-x^2/(4y)} \, dy$$

$$- \int_t^\infty f(t_0) \frac{x}{2\sqrt{\pi y^3}} e^{-x^2/(4y)} \, dy = I_1 + I_2 + I_3,$$

where I_1, I_2, I_3 denote the integrals on the intervals $(0, \delta)$, (δ, t), (t, ∞) respectively. Suppose that $|t - t_0| < \delta$. Then, by formula (9) of Example 4.4.12,

$$|I_1| < \frac{\varepsilon}{3} \int_0^\infty \frac{x}{2\sqrt{\pi y^3}} \, e^{-x^2/(4y)} \, dy = \frac{\varepsilon}{3}.$$

There is a number $\eta_0 > 0$ such that for $0 < x < \eta_0$

$$|I_2| < \frac{4M}{\sqrt{\pi}} \int_{x/\sqrt{4t}}^{x/\sqrt{4\delta}} e^{-z^2} \, dz < \frac{4M}{\sqrt{\pi}} \int_0^{x/\sqrt{4\delta}} e^{-z^2} \, dz < \frac{\varepsilon}{3}.$$

Further there is a number η, $0 < \eta \leqslant \eta_0$, such that for $0 < x < \eta$

$$|I_3| < \int_{t_0-\delta}^\infty -M \frac{x}{2\sqrt{\pi y^3}} \, e^{-x^2/(4y)} \, dy = \frac{2M}{\sqrt{\pi}} \int_0^{x/\sqrt{4(t_0-\delta)}} e^{-z^2} \, dz < \frac{\varepsilon}{3}.$$

Then, for $|t - t_0| < \delta$ and $0 < x < \eta$,

$$|u(x, t) - f(t_0)| < \varepsilon.$$

It follows that $u(x, t)$ has property (6c) and also in the remaining case $(x = 0, t > 0)$ property (6d). Hence (7) is a solution of the problem.

3. Suppose that problem (6) has two solutions. Let $u(x,t)$ be their difference. Then $u(x,t)$ satisfies properties (6) with the exception that (6c) is replaced by $u(0, t) = 0, t > 0$. Suppose that M is a positive number such that $|u(x,t)| < M$ for $x > 0$ and $t > 0$. Suppose that $a > 0$. Then the values of the function $u(x,t) - Mx/a$ are negative or zero on the boundary of the region $0 < x < a, t > 0$. Further this function is a solution of the heat equation in the quadrant $x > 0, t > 0$. The maximum principle then shows that

$$u(x, t) - Mx/a \leqslant 0 \quad \text{for} \quad 0 < x < a, \quad t > 0.$$

Fix the point (x, t), and let a tend to infinity. It follows that $u(x, t) \leqslant 0$. Hence the function $u(x, t)$ does not take on positive values. Application of this result to the function $-u(x, t)$ shows that $u(x, t)$ does not take on negative values. It follows that the function (7) is the only solution of the problem.

4. In the same way as in part 2 of the solution of Problem 8.4.3 it can be shown that the solution (7) is stable (Exercise 832).

Then problem (6) has the solution (7), and the problem is correctly posed.

The next problem has the following physical interpretation. Suppose that the heat distribution in a thin circular uniform plate is known at the time $t = 0$ (this distribution is so chosen that the computations become fairly short), that the boundary of the plate is always kept at zero temperature, and that the plate has no exchange of heat with its surroundings except at its boundary. Find the distribution of heat on the plate at the time $t > 0$.

8.4.6 Problem

Solve the boundary-value problem

$$\frac{\partial u}{\partial t} = \frac{\partial^2 u}{\partial x^2} + \frac{\partial^2 u}{\partial y^2}, \quad x^2 + y^2 < 1, \quad t > 0, \tag{9a}$$

$$u(\cos \theta, \sin \theta, t) = 0, \qquad \theta \in R, \quad t > 0, \tag{9b}$$

$$u(r \cos \theta, r \sin \theta, 0) = J_0(j_1 r), \qquad 0 < r < 1, \quad \theta \in R, \tag{9c}$$

$$u(x, y, t) \quad \text{is continuous for} \quad x^2 + y^2 \leqslant 1, \qquad t \geqslant 0, \tag{9d}$$

where J_0 is the Bessel function of order zero, and where j_1 is the smallest positive zero of this function (cf. Section 6.3.1 and Theorem 6.5.2). Show that the problem is correctly posed.

Solution. **1.** We change to polar coordinates r and θ, and we assume that the solution is independent of θ. Then the following boundary-value problem is obtained (Exercise 838):

$$\frac{\partial u}{\partial t} = \frac{\partial^2 u}{\partial r^2} + \frac{1}{r}\frac{\partial u}{\partial r}, \qquad 0 < r < 1, \quad t > 0, \tag{10a}$$

$$u(1, t) = 0, \qquad\qquad\qquad t > 0, \tag{10b}$$

$$u(r, 0) = J_0(j_1 r), \qquad\qquad 0 < r < 1, \tag{10c}$$

$$u(r, t) \quad \text{is continuous for} \quad 0 \leqslant r \leqslant 1, \qquad t \geqslant 0. \tag{10d}$$

We argue as in part **1** of the solution of Problem 8.3.4. We write down the formulas and leave it to the reader to fill in accompanying comments:

$$u(r, t) = R(r)T(t), \qquad RT' = R''T + \frac{1}{r}R'T,$$

$$\frac{T'}{T} = \frac{R''}{R} + \frac{1}{r}\frac{R'}{R} = -\lambda, \qquad r^2 R'' + rR' + \lambda r^2 R = 0,$$

$$\lambda = j_1^2, \qquad R(r) = J_0(j_1 r), \qquad T(t) = e^{-j_1^2 t},$$

$$u(r, t) = J_0(j_1 r)e^{-j_1^2 t}. \tag{11}$$

2. It can be verified (Exercise 839), that the function (11) satisfies the boundary-value problem (10), and that the function

$$u(x, y, t) = J_0(j_1 \sqrt{x^2 + y^2})e^{-j_1^2 t} \tag{12}$$

satisfies the boundary-value problem (9).

3. It can be verified (Exercise 839) that the maximum principle has a counterpart for the heat equation (9a) on the closed set $x^2 + y^2 \leqslant 1, t \geqslant 0$, and hence that the solution (12) is unique and stable.

Then problem (10) has the solution (12), and the problem is correctly posed.

EXERCISES

828. Show that the function (2) satisfies condition (1a).

829. Find a solution $u(x, t)$ of the boundary-value problem

$$u_t' = u_{xx}'', \qquad\qquad 0 < x < \pi, \quad t > 0,$$
$$u(x, 0) = x \sin x, \qquad 0 < x < \pi,$$
$$u(0, t) = u(\pi, t) = 0, \qquad t > 0.$$

830. Find a solution $u(x, t)$ of the boundary-value problem

$$u_t' = u_{xx}'', \qquad\qquad 0 < x < \pi, \quad t > 0,$$
$$u(x, 0) = \sin^2 x, \qquad 0 < x < \pi,$$
$$u_x'(0, t) = u_x'(\pi, t) = 0. \qquad t > 0.$$

*831. Show that the function (7) has properties (6a), (6b) and (6e), and that it is continuous for $x > 0, t \geqslant 0$, and for $x = t = 0$.

832. Show that the solution (7) of Problem 8.4.5 is stable.

833. Find a bounded solution $u(x, t)$ of the boundary-value problem

$$u_t' = u_{xx}'', \qquad x \in R, \quad t \geqslant 0,$$
$$u(x, 0) = e^{-x^2}, \qquad x \in R.$$

Is the problem correctly posed?

834. Suppose that $f(x, y)$ is a bounded and continuous function with domain R^2. Show that the function

$$u(x, y, t) = \frac{1}{4\pi t} \iint_{R^2} f(\xi, \eta) \exp\left[-\frac{(x - \xi)^2 + (y - \eta)^2}{4t}\right] d\xi \, d\eta$$

satisfies the heat equation

$$u_t' = u_{xx}'' + u_{yy}'', \qquad (x, y) \in R^2, \quad t > 0.$$

Show that

$$\lim_{t \to 0} u(x, y, t) = f(x, y).$$

835. Find a solution $u(x, t)$ of the boundary-value problem

$$u_t' = u_{xx}'', \qquad x > 0, \quad t > 0,$$
$$u(x, 0) = 0, \qquad x > 0,$$
$$u(0, t) = 1, \qquad t > 0.$$

Is the problem correctly posed?

836. Show that if the function $u(x, t)$ has the domain $0 \leqslant x \leqslant 1, 0 \leqslant t \leqslant 1$, and if

$$u_t' = u_{xx}'', \qquad 0 \leqslant x \leqslant 1, \quad 0 \leqslant t \leqslant 1,$$
$$u_x'(0, t) = 0, \qquad 0 \leqslant t \leqslant 1,$$

then $u(x, t)$ takes its largest value on the straight line $t = 0$ or on the straight line $x = 1$.

837. The boundary-value problem

$$u_t' = u_{xx}'', \qquad x > 0, \quad t > 0,$$
$$u(x, 0) = \begin{cases} 1 & \text{for } 0 < x < 1 \\ 0 & \text{for } 1 < x, \end{cases}$$

$$u'_x(0, t) = 0, \quad t > 0$$

has a solution of the form

$$u(x, t) = \int_0^\infty f(y, t) \cos xy \, dy.$$

Find this solution.

Hint. Use the cosine transform of $u(x, 0)$.

838. Show that the change of coordinates $x = r \cos \theta$, $y = r \sin \theta$, and the assumption that the solution is independent of θ transforms problem (9) into problem (10).

839. Carry through the verifications in parts **2** and **3** of the solution of Problem 8.4.6.

840. Consider the boundary-value problem that is obtained from (9) by exchanging (9c) for $u(r \cos \theta, r \sin \theta, 0) = f(r)$, $0 < r < 1$, $\theta \in R$, where $f(r)$ is a given function continuous on the interval $[0, 1]$ and where $f(1) = 0$. Suppose that $\varepsilon > 0$, and that we have determined a function $g(r)$ of the form $g(r) = \sum_{v=1}^n c_v J_0(j_v r)$, where the c_v denote constants and where the j_v denote the n smallest positive zeros of the Bessel function J_0, such that $|g(r) - f(r)| < \varepsilon$ for $0 < r < 1$. Show that the problem is correctly posed. Find an approximation $v(x, y, t)$ of the solution $u(x, y, t)$ of the problem. Estimate the error $v(x, y, t) - u(x, y, t)$ for $x^2 + y^2 < 1, t > 0$.

8.5 THE LAPLACE EQUATION

The following theorem is a counterpart to the Laplace equation of Theorem 8.4.2 on the heat equation.

8.5.1 Theorem (the maximum principle for the Laplace equation). *Suppose that Ω is a bounded region in the xy-plane, that $\partial\Omega$ is the boundary of Ω, and that Ω^- is the union of Ω and $\partial\Omega$. Suppose that $u(x, y)$ is a continuous function with domain Ω^-, and that $u(x, y)$ satisfies the Laplace equation $u''_{xx} + u''_{yy} = 0$ for $(x, y) \in \Omega$. Then $u(x, y)$ takes its largest value, and also its smallest value, on $\partial\Omega$.*

Proof. Suppose $\varepsilon > 0$. Set

$$v(x, y) = u(x, y) + \varepsilon(x^2 + y^2).$$

Then

(a) $$\frac{\partial^2 v}{\partial x^2} + \frac{\partial^2 v}{\partial y^2} = 4\varepsilon \qquad \text{for} \quad (x, y) \in \Omega.$$

Suppose that $(x_0, y_0) \in \Omega^-$ and that
$$v(x_0, y_0) = \max_{\Omega^-} v(x, y).$$

It is seen that

$$(x_0, y_0) \in \Omega \;\Rightarrow\; \frac{\partial^2 v}{\partial x^2}(x_0, y_0) \leqslant 0 \qquad \text{and} \qquad \frac{\partial^2 v}{\partial y^2}(x_0, y_0) \leqslant 0.$$

These inequalities contradict (a). Hence $(x_0, y_0) \in \partial\Omega$. Set $M = \max\limits_{\partial\Omega} u(x, y)$. Then

$$\max_{\Omega^-} u(x, y) \leqslant \max_{\Omega^-} v(x, y) = \max_{\partial\Omega} v(x, y) \leqslant M + \varepsilon \max_{\partial\Omega} (x^2 + y^2),$$

$$\max_{\Omega^-} u(x, y) \leqslant M, \qquad \max_{\Omega^-} u(x, y) = M,$$

which is the first assertion. The second assertion is obtained by applying the first assertion to the function $-u(x, y)$.

8.5.2 Definition

Let Ω and $\partial\Omega$ have the same meaning as in Theorem 8.5.1. The *Dirichlet problem* for the domain Ω and the boundary function u_0 is the boundary-value problem:

$$\frac{\partial^2 u}{\partial x^2} + \frac{\partial^2 u}{\partial y^2} = 0, \qquad (x, y) \in \Omega, \tag{1a}$$

$$u(x, y) = u_0(x, y), \qquad (x, y) \in \partial\Omega, \tag{1b}$$

where u_0 is a given function with domain $\partial\Omega$.

The Dirichlet problem has the following physical interpretation. Find the temperature at every point of a uniform plate Ω when the temperature at each point of its boundary is known, and there is no exchange of heat with the surroundings except at the boundary, and the temperature distribution on the plate is stationary, i.e. it does not change with time. Below we solve the Dirichlet problem in the case where Ω is a circular disk. This case has historic interest: Fourier's work on this problem was the beginning of Fourier analysis.

8.5.3 Problem

Solve the boundary-value problem

$$\frac{\partial^2 u}{\partial x^2} + \frac{\partial^2 u}{\partial y^2} = 0, \qquad x^2 + y^2 < 1, \tag{2a}$$

$$u(\cos\theta, \sin\theta) = f(\theta), \qquad \theta \in R, \tag{2b}$$

$$u(x, y) \quad \text{is continuous for} \quad x^2 + y^2 \leqslant 1, \tag{2c}$$

where $f(\theta)$ is a given function in the class $C^2(R)$, periodic with period 2π. Show that the problem is correctly posed.

Solution. 1. Changing to polar coordinates r and θ (Exercise 844) gives the boundary-value problem

$$\frac{\partial^2 u}{\partial r^2} + \frac{1}{r}\frac{\partial u}{\partial r} + \frac{1}{r^2}\frac{\partial^2 u}{\partial \theta^2} = 0, \qquad 0 < r < 1, \quad \theta \in R, \tag{3a}$$

$$u(1, \theta) = f(\theta), \qquad \theta \in R, \tag{3b}$$

$$u(0, \theta) \quad \text{is constant for} \quad \theta \in R, \tag{3c}$$

$$u(r, \theta + 2\pi) = u(r, \theta), \qquad 0 < r < 1, \quad \theta \in R, \tag{3d}$$

$$u(r, \theta) \text{ is continuous for } \quad 0 \leqslant r \leqslant 1, \qquad \theta \in R. \tag{3e}$$

We argue as in part **1** of the solution of Problem 8.3.4. We write down the formulas and leave it to the reader to fill in accompanying comments:

$$v(r, \theta) = R(r)\,\Theta\,(\theta), \qquad R''\Theta + \frac{1}{r}\,R'\Theta + \frac{1}{r^2}\,R\Theta'' = 0,$$

$$r^2\,\frac{R''}{R} + r\,\frac{R'}{R} = -\frac{\Theta''}{\Theta} = \lambda,$$

$$\Theta'' + \lambda\Theta = 0, \qquad \lambda = n^2, \quad n \in N,$$

$$\Theta(\theta) = \tfrac{1}{2}a_0 \qquad \text{if} \quad n = 0,$$

$$\Theta(\theta) = a_n \cos n\theta + b_n \sin n\theta \qquad \text{if} \quad n > 0,$$

$$r^2 R'' + rR' - n^2 R = 0,$$

$$n = 0 \Rightarrow R = c_1 + c_2 \log r, \qquad c_2 = 0, \quad c_1 = 1,$$

$$n > 0 \Rightarrow R = c_1 r^n + c_2 r^{-n}, \qquad c_2 = 0, \quad c_1 = 1,$$

$$v(r, \theta) = \tfrac{1}{2}a_0 \qquad \text{if} \quad n = 0,$$

$$v(r, \theta) = r^n(a_n \cos n\theta + b_n \sin n\theta) \qquad \text{if} \quad n > 0,$$

$$f(\theta) \sim \tfrac{1}{2}a_0 + \sum_{n=1}^{\infty} (a_n \cos n\theta + b_n \sin n\theta),$$

$$u(r, \theta) = \tfrac{1}{2}a_0 + \sum_{n=1}^{\infty} r^n(a_n \cos n\theta + b_n \sin n\theta). \tag{4}$$

2. The function (4) satisfies the boundary-value problem (3) and also the boundary-value problem (2) (Exercise 845).

3. Suppose that problem (2) has two solutions. Let $u(x, t)$ be their difference. Then $u(x, t)$ satisfies conditions (2) with the exception that (2b) is replaced by $u(\cos \theta, \sin \theta) = 0$, $\theta \in R$. The maximum principle shows that $u(x, y) = 0$ for $x^2 + y^2 \leqslant 1$. Then (4) is the only solution of the problem.

4. By use of the maximum principle it can be shown that the solution (4) is stable (Exercise 846).

Then problem (2) has the solution (4), and the problem is correctly posed.

8.5.4 Remark

Suppose that $0 \leqslant r < 1$. The series in (4) can then be replaced by the integral in (5) below, called the *Poisson integral*. Hence this integral furnishes an alternative expression for the solution of the Dirichlet problem (2). This is seen by the following computations:

$$u(r, \theta) = \frac{1}{2\pi} \int_{-\pi}^{\pi} f(t)\,dt$$

$$+ \frac{1}{\pi} \sum_{n=1}^{\infty} r^n [\cos n\theta \int_{-\pi}^{\pi} f(t) \cos nt \, dt$$

$$+ \sin n\theta \int_{-\pi}^{\pi} f(t) \sin nt \, dt]$$

$$= \frac{1}{2\pi} \int_{-\pi}^{\pi} f(t) \left[1 + 2 \sum_{n=1}^{\infty} r^n \cos n(\theta - t) \right] dt,$$

$$1 + 2 \sum_{n=1}^{\infty} r^n \cos n\theta = 1 + 2 \operatorname{Re} \sum_{n=1}^{\infty} r^n e^{in\theta} = 1 + 2 \operatorname{Re} \frac{re^{i\theta}}{1 - re^{i\theta}}$$

$$= 1 + \frac{re^{i\theta}}{1 - re^{i\theta}} + \frac{re^{-i\theta}}{1 - re^{-i\theta}} = \frac{1 - r^2}{1 - 2r \cos \theta + r^2},$$

$$u(r, \theta) = \frac{1 - r^2}{2\pi} \int_{-\pi}^{\pi} \frac{f(t)}{1 - 2r \cos (\theta - t) + r^2} \, dt. \tag{5}$$

8.5.5 Remark

A function $u(x, y)$ that satisfies the Laplace equation is called a *harmonic function*. Suppose that U is the unit disk $x^2 + y^2 < 1$ in the xy-plane, that Ω is a region in the $\xi\eta$-plane, and that the functions

$$x = x(\xi, \eta), \qquad y = y(\xi, \eta), \tag{6}$$

define a conformal mapping of Ω onto U (i.e. they belong to the class $C^2(\Omega)$, and they map Ω one to one onto U so that magnitude and direction of angles are preserved; for conformal mapping see [1], p. 73 *et seq.*, p. 89 *et seq.*, and Ch. 6). It can be verified that if the function $u(x, y)$ is harmonic on U, then the composite function $u(x(\xi, \eta), y(\xi, \eta))$ is harmonic on Ω (Exercise 847). Hence, if we know a conformal mapping (6) of Ω onto U, we can solve the Dirichlet problem for the region Ω and certain boundary functions $u_0(\xi, \eta)$.

EXERCISES

In Exercises 841–843 and 848–851 we say "*a* solution", not "*the* solution". Exercise 852 demonstrates the reason for this.

841. Find a solution $u(x, y)$ of the boundary-value problem

$$
\begin{aligned}
u''_{xx} + u''_{yy} &= 0, & 0 < x < \pi, \quad 0 < y < 1, \\
u(x, 0) &= 0, & 0 < x < \pi, \\
u(0, y) = u(\pi, y) &= 0, & 0 < y < 1, \\
u(x, 1) &= \begin{cases} x & \text{for} \quad 0 < x < \pi/2 \\ \pi - x & \text{for} \quad \pi/2 < x < \pi. \end{cases}
\end{aligned}
$$

842. Find a bounded solution $u(r, \theta)$ of the boundary-value problem

$$r^2 u''_{rr} + ru'_r + u''_{\theta\theta} = 0, \qquad 1 < r, \quad -\pi < \theta < \pi,$$
$$u(1, \theta) = \theta, \qquad\qquad\qquad -\pi < \theta < \pi,$$
$$u(r, -\pi) = u(r, \pi) = 0, \qquad 1 < r.$$

843. Find a solution $u(r, \theta)$ of the boundary-value problem

$$r^2 u''_{rr} + ru'_r + u''_{\theta\theta} = 0, \qquad 0 < r < 1, \quad 0 < \theta < \pi,$$
$$u(1, \theta) = \theta(\pi - \theta), \qquad 0 < \theta < \pi,$$
$$u(0, \theta) = 0, \qquad\qquad 0 < \theta < \pi,$$
$$u(r, 0) = u(r, \pi) = 0, \qquad 0 < r < 1.$$

*844. Show that the change of coordinates $x = r \cos \theta$, $y = r \sin \theta$, transforms problem (2) into problem (3).

845. Show that the function (4) satisfies problems (3) and (2).

846. Show that the solution (4) of problem (2) is stable.

847. Prove the claim in the comment to (6).

848. Show that the Laplace equation $\Delta_3 u = 0$ in spherical coordinates (θ is measured along the "equator" and ϕ from the "north pole") can be written

$$\frac{\partial}{\partial r}\left(r^2 \frac{\partial u}{\partial r}\right) + \frac{1}{\sin \phi} \frac{\partial}{\partial \phi}\left(\sin \phi \frac{\partial u}{\partial \phi}\right) = 0$$

for solutions that are independent of θ. Find a solution $u(r, \phi)$ in the spherical region $x^2 + y^2 + z^2 < 1$ that is independent of θ and satisfies the boundary condition
$$u(1, \phi) = \cos 3\phi, \qquad 0 < \phi < \pi.$$
Hint. Set $t = \cos \phi$.

849. Suppose that f_0, f_1, g_0, g_1 are given functions, continuous on the interval $[0, 1]$. Find for the boundary-value problem

$$u''_{xx} + u''_{yy} = 0, \qquad 0 < x < 1, \quad 0 < y < 1,$$
$$u(x, 0) = f_0(x), \qquad 0 < x < 1,$$
$$u(x, 1) = f_1(x), \qquad 0 < x < 1,$$
$$u(0, y) = g_0(y), \qquad 0 < y < 1,$$
$$u(1, y) = g_1(y), \qquad 0 < y < 1,$$

a solution $u(x, y)$ that is represented as the sum of four functions, each vanishing on three sides of the square.

850. The boundary-value problem

$$u''_{xx} + u''_{yy} = 0, \qquad x > 0, \quad 0 < y < 1,$$
$$u(x, 0) = 0, \qquad\qquad x > 0,$$
$$u(0, y) = 0, \qquad\qquad 0 < y < 1,$$
$$u(x, 1) = xe^{-x}, \qquad x > 0,$$

has a solution of the form

$$u(x, y) = \int_0^\infty f(t, y) \sin tx \, dt.$$

Find this solution.

Hint. Use the sine transform of $u(x, 1)$.

851. Consider the following boundary-value problem in the $\xi\eta$-plane:

$$u_{\xi\xi}'' + u_{\eta\eta}'' = 0, \qquad \xi > 0, \eta \in R,$$
$$u(0, \eta) \quad = \begin{cases} 1 & \text{for} \quad |\eta| < 1 \\ 0 & \text{for} \quad |\eta| > 1. \end{cases}$$

a) Give a physical interpretation of the problem.
b) Find a conformal mapping that carries the half-plane $\xi > 0$ onto the unit disk $x^2 + y^2 < 1$ of the xy-plane, so that the origin is carried into the point $(1, 0)$, and the point $(1, 0)$ into the origin.
c) This mapping produces a boundary-value problem of type (2). Find a solution of the form (4), and also of the form (5), for the new problem.
d) At which points (ξ, η) does the corresponding solution of the first problem take the value $\frac{1}{2}$?

852. Is the boundary-value problem

$$u_{xx}'' + u_{yy}'' \quad = 0, \qquad x^2 + y^2 < 1,$$
$$u(\cos\theta, \sin\theta) = 0, \qquad 0 < \theta < 2\pi,$$
$$u(x, y) \quad \text{is continuous if} \quad x^2 + y^2 \leqslant 1 \quad \text{and} \quad x < 1,$$

correctly posed?

Hint. Let c be a real number. Is the real part of the function $f(z) = c(z + 1)/(z - 1)$, $z = x + iy$, a solution of the problem?

ANSWERS TO EXERCISES

CHAPTER 1

101. (a) and (b) are integrable; (c), (d) and (e) are not.

103. No. (It can in fact be proved that the sum $f(x)$ of the series is discontinuous at every multiple of 2π. It can be shown that $f(0) = 0$ and that $f(x) = (\pi - x)/2$ for $0 < x < 2\pi$; cf. Exercise 116 and Theorem 1.4.4.)

104. a) $\sum_{n=1}^{\infty}(2n - 1)^{-3}\sin(2n - 1)x = (\pi^2 x - \pi x^2)/8$ for $0 \leqslant x \leqslant \pi$,
b) $\sum_{n=1}^{\infty}(2n - 1)^{-3}\sin(2n - 1)x = (\pi x^2 - 3\pi^2 x + 2\pi^3)/8$ for $\pi \leqslant x \leqslant 2\pi$.

105. $a > 1$.

106. a) $x \sim 2\sum_{n=1}^{\infty}(-1)^{n+1}\dfrac{\sin nx}{n}$ for $-\pi < x < \pi$,

b) $x \sim \pi - 2\sum_{n=1}^{\infty}\dfrac{\sin nx}{n}$ for $0 < x < 2\pi$.

107. $2/\pi$ and 0 respectively.

108. $0 < a < 1$.

110. a) No x, b) $x = k\pi, k \in Z$.

111. a) Yes, b) no, c) no.

112. No.

113. Yes.

114. $a = \pi^2/3, b = -4, c = 0$.

116. $f(x) \sim \sum_{n=1}^{\infty}\dfrac{\sin nx}{n}$.

117. a) $f(x) \sim \dfrac{1}{4} + \sum_{n=1}^{\infty}\dfrac{(-1)^{n+1}}{n\pi}\left[\dfrac{(-1)^n - 1}{n\pi}\cos nx + \sin nx\right]$,

b) $\pi^2/8$.

118. a) $f(x) \sim 4\pi^2/3 + 4\sum_{n=1}^{\infty}(\cos nx - n\pi\sin nx)/n^2$,
b) $\pi^2/6$.

119. $f(x) \sim \dfrac{\sinh \pi}{\pi}\left[1 + 2\sum_{n=1}^{\infty}(-1)^n\dfrac{\cos nx}{1 + n^2}\right]$.

122. b) Yes.

123. $(1 + \cos 2x)/2; \displaystyle\int_{-\pi}^{\pi}(1 + \tan^2 x)^{-2}\,dx = 3\pi/4$.

124. $\dfrac{\pi}{2} - \dfrac{4}{\pi} \displaystyle\sum_{n=1}^{\infty} (2n-1)^{-2} \cos(2n-1)x; \pi^4/96.$

125. $12\displaystyle\sum_{n=1}^{\infty}(-1)^n n^{-3} \sin nx; \pi^6/945.$

128. $c \sin x$, where c is a real number.

130. $e^{\cos x}\cos(\sin x)$ and $e^{\cos x}\sin(\sin x)$.

131. $\log[2\cos(x/2)]$ and $x/2$.

CHAPTER 2

201. a) $\frac{2}{3} + i$, b) 30, c) $1 + i$, d) $(2-i)(1-e^{-4\pi})/5$.

202. Each zero of $P(x)$ is nonreal.

207. d) $\dfrac{1}{\sqrt{\pi}}, \sqrt{\dfrac{2}{\pi}}\cos x, \sqrt{\dfrac{2}{\pi}}\cos 2x, \ldots, \sqrt{\dfrac{2}{\pi}}\cos nx, \ldots.$

208. a) $(e^\pi - 1)/\pi + \dfrac{2}{\pi}\displaystyle\sum_{n=1}^{\infty}(1+n^2)^{-1}[(-1)^n e^\pi - 1]\cos nx,$

   c) $\dfrac{2}{\pi}\displaystyle\sum_{n=1}^{\infty} n(1+n^2)^{-1}[1-(-1)^n e^\pi]\sin nx.$

210. $\displaystyle\sum_{n=-\infty}^{\infty} c_n e^{inx}$, where $c_0 = 0$ and $c_n = -i/n$ for $n \neq 0$.

213. a) $\dfrac{128}{\pi^3}\displaystyle\sum_{n=1}^{\infty}(2n-1)^{-3}\sin\dfrac{(2n-1)\pi x}{4},$

   b) $\dfrac{8}{3} - \dfrac{16}{\pi^2}\displaystyle\sum_{n=1}^{\infty} n^{-2}\cos\dfrac{n\pi x}{2}.$

214. $\dfrac{1}{\pi} - \dfrac{4}{\pi}\displaystyle\sum_{n=1}^{\infty}(4n^2-1)^{-1}\cos\dfrac{2n\pi x}{a}.$

215. b) $b_{2n-1} = \dfrac{4}{\pi}\displaystyle\int_0^{\pi/2} f(x)\sin(2n-1)x\, dx, n \in Z^+.$

219. Yes.

220. Yes.

221. Yes.

222. $n = 6$.

223. $f(x) \sim \displaystyle\sum_{n=-\infty}^{\infty} \dfrac{\sin(n-a)\pi}{(n-a)\pi} e^{inx}.$

225. Yes.

226. No. (The function $\cos 2\pi x$ is orthogonal on the interval $(0, 1)$ to every function of the system.)

227. $(3x^2 - 6\pi x + 2\pi^2)/12, -\displaystyle\int_0^x \log\left(2\sin\dfrac{t}{2}\right)dt.$

228. $\pi\sinh\pi + \dfrac{4\sinh\pi}{\pi}\displaystyle\sum_{n=1}^{\infty}(2n-1)^{-2}(2n^2-2n+1)^{-1}.$

229. There exist two complex numbers c_1 and c_2, not both zero, such that $c_1 f(x) = c_2 g(x)$ on (a, b).

232. a) Yes, b) no, c) yes, d) no.

233. $\cos nx$, $n \in N$, $x \in [0, \pi]$.

234. No. (Set $\phi_n(\pi/2) = 1$ in Exercise 233 to get a counterexample.)

235. $\phi_0(x) = 1$, $\phi_1(x) = x$, $\phi_2(x) = x^2 - \frac{1}{3}$, $\phi_3(x) = x^3 - \frac{3}{5}x$.

238. a) $\lambda_n = n^2$, $\phi_n(x) = c_n \cos nx$, $n \in N$, $c_n \neq 0$,
   b) $\lambda_n = n^2\pi^2$, $\phi_n(x) = c_n \sin n\pi x$, $n \in Z^+$, $c_n \neq 0$,
   c) $\lambda_n = (2n + 1)^2$, $\phi_n(x) = c_n \sin (2n + 1)x$, $n \in N$, $c_n \neq 0$.

239. a) $\lambda_n = n^2$, $\phi_n(x) = c_n e^{-x} \sin nx$, $n \in Z^+$, $c_n \neq 0$.
   b) The eigenvalues are the positive roots of the equation $\tan \sqrt{\lambda} + \sqrt{\lambda} = 0$. (Let them be arranged in an increasing sequence λ_n, $n \in Z^+$. Sketch the curves $y = \tan x$ and $y = -x$. It is seen that λ_n is a little larger than $(n\pi - \pi/2)^2$.) The eigenfunctions are $\phi_n(x) = c_n \sin \sqrt{\lambda_n}x$, $c_n \neq 0$.

CHAPTER 3

302. $P_2(x)$ has a minimum $= -\frac{1}{2}$ for $x = 0$. $P_3(x)$ has a maximum $= 5^{-1/2} \approx 0{\cdot}45$ for $x = -5^{-1/2}$ and a minimum $= -5^{-1/2}$ for $x = 5^{-1/2}$.

304. a) $P_{2n}(0) = (-1)^n 2^{-2n} \binom{2n}{n} = (-1)^n 2^{-n}(n!)^{-1}(2n - 1)!!$, $P_{2n+1}(0) = 0$, $n \in N$.
   b) Yes, the limit is 0.

305. $P_n(x) = (b - a)^{-n}(n!)^{-1}d^n[(x - a)^n(x - b)^n]/dx^n$.

307. $f(x) \sim \frac{1}{3}P_0(x) + \frac{2}{3}P_2(x)$.

308. a) $\frac{1}{2}P_0(x) + \frac{5}{8}P_2(x)$, b) $\frac{15}{16}x^2 + \frac{3}{16}$.

309. $\sqrt{\dfrac{2n + 1}{2}}\, P_n(x)$, $n \in N$, $x \in (-1, 1)$.

311. $p_0(x) = 1$, $p_1(x) = \sqrt{3}(2x - 1)$, $p_2(x) = \sqrt{5}(6x^2 - 6x + 1)$.

314. Yes.

315. $35x^4 - 30x^2 + 3 \,(= 8P_4(x))$.

318. $\frac{1}{2} \log \dfrac{1 + x}{1 - x}$.

320. 2^{n-1}.

321. $(x^2 - 1)^{n/2}$ if n is even, $nx(x^2 - 1)^{(n-1)/2}$ if n is odd.

322. 0 if n is odd, $-2(n^2 - 1)^{-1}$ if n is even.

325. $x = \cos \dfrac{v\pi}{n}$, $v \in \{1, 2, \ldots, n - 1\}$.

328. a) $\dfrac{d}{dx}((1 - x^2)^{1/2}y') + \lambda(1 - x^2)^{-1/2}y = 0$.
   b) $\lim_{x \to 1} y(x)$ and $\lim_{x \to -1} y(x)$ exist.
   c) $\lambda_n = n^2$, $n \in N$.
   d) $T_n(x)$, $n \in N$.

329. a) $f(x) = \frac{3}{2} T_0(x) + T_1(x) + \frac{1}{2} T_2(x)$,
   b) $f(x) \sim \frac{3}{2} T_0(x) + T_1(x) + \frac{1}{2} T_2(x)$.

330. $\dfrac{\pi}{2} - \dfrac{4}{\pi} \displaystyle\sum_{n=1}^{\infty} (2n-1)^{-2} T_{2n-1}(x).$

CHAPTER 4

402. a) $1 < \alpha < 2$, b) $0 < \alpha < 2$.

403. a) No, b) yes, c) no.

404. a) No, b) yes.

405. 0 and log 2.

406. $a(t) = \dfrac{1}{\pi} \displaystyle\int_{-\infty}^{\infty} f(u) \cos tu\, du,\ b(t) = \dfrac{1}{\pi} \displaystyle\int_{-\infty}^{\infty} f(u) \sin tu\, du.$

408. No.

410. a) $F(t) = \dfrac{2}{\pi} t^{-2} \sin^2 \dfrac{t}{2}$ for $t \neq 0$, $F(0) = \dfrac{1}{2\pi}$,

b) $\displaystyle\int_{-\infty}^{\infty} \dfrac{2}{\pi} t^{-2} \sin^2 \dfrac{t}{2}\, e^{itx}\, dx = \begin{cases} 1 - |x| & \text{for } |x| \leqslant 1, \\ 0 & \text{for } |x| > 1. \end{cases}$

411. a) $F_c(t) = \dfrac{2}{\pi} \dfrac{\cos (\pi t/2)}{1 - t^2}$ for $0 \leqslant t < 1$ and $1 < t$, $F_c(1) = \tfrac{1}{2}$,

b) $\dfrac{\pi}{2}$.

412. a) $F_s(t) = \dfrac{1}{\pi} \dfrac{4t}{t^4 + 4}$, $t \geqslant 0$, b) $\dfrac{\pi \sin 1}{e}$.

413. a) $\dfrac{e^{-|t|}}{2}$, b) $\tfrac{1}{4} e^{-|t|} (1 + |t|)$,

c) $\dfrac{\cos t + i \sin t}{6} (e^{-|t|} - \tfrac{1}{2} e^{-2|t|}).$

415. No. (The Fourier transform of $f(x)$ is not integrable on $(-\infty, \infty)$.)

417. a) $\alpha F(t) + \beta G(t)$, b) $F(t - a)$, c) $\dfrac{1}{|a|} F\left(\dfrac{t}{a}\right).$

422. $F(t) = \dfrac{1}{\pi} \dfrac{\sin t}{t}$ for $t \neq 0$, $\quad F(0) = \dfrac{1}{\pi}$;

$(f * f)(x) = \begin{cases} 2 - |x| & \text{for } |x| \leqslant 2, \\ 0 & \text{for } |x| > 2. \end{cases}$

423. $\displaystyle\int_{-\infty}^{\infty} t^{-4} (t \cos t - \sin t)^2\, dt = \dfrac{\pi}{3}.$

424. $\displaystyle\int_{0}^{\infty} |F_c(t)|^2\, dt = \dfrac{2}{\pi} \displaystyle\int_{0}^{\infty} |f(x)|^2\, dx,$

$\displaystyle\int_{0}^{\infty} |F_s(t)|^2\, dt = \dfrac{2}{\pi} \displaystyle\int_{0}^{\infty} |f(x)|^2\, dx.$

CHAPTER 5

501. a) $\dfrac{1}{s-2}$, $\operatorname{Re}(s) > 2$ b) $\dfrac{1}{s+2}$, $\operatorname{Re}(s) > -2$ c) $\dfrac{1}{s-i}$, $\operatorname{Re}(s) > 0$

d) $\dfrac{1}{s+i}$, $\operatorname{Re}(s) > 0$ e) $\dfrac{1}{s}$, $\operatorname{Re}(s) > 0$ f) $\dfrac{1}{s^2}$, $\operatorname{Re}(s) > 0$

g) $\dfrac{2}{s^3}$, $\operatorname{Re}(s) > 0$ h) $\dfrac{1}{s} - \dfrac{2}{s^3}$, $\operatorname{Re}(s) > 0$ i) $\dfrac{s}{s^2+1}$, $\operatorname{Re}(s) > 0$

j) $\dfrac{s}{s^2+4}$, $\operatorname{Re}(s) > 0$ k) $\dfrac{1}{s^2+1}$, $\operatorname{Re}(s) > 0$ l) $\dfrac{2}{s^2+4}$, $\operatorname{Re}(s) > 0$

m) $\dfrac{s}{s^2-1}$, $\operatorname{Re}(s) > 1$ n) $\dfrac{s}{s^2-4}$, $\operatorname{Re}(s) > 2$ o) $\dfrac{1}{s^2-1}$, $\operatorname{Re}(s) > 1$

502. a) $\operatorname{Re}(s) > 0$, b) $\operatorname{Re}(s) \geqslant 0$,
c) the s-plane, d) the empty set.

503. The real and imaginary parts of $(n!)\left(\dfrac{\alpha + i\beta}{\alpha^2 + \beta^2}\right)^{n+1}$ respectively.

504. $f_1(t) = e^{at}$ for $t > 0$, $f_1(0) = \frac{1}{2}$, $f_1(t) = 0$ for $t < 0$.

506. a) te^{at}, b) $\sin at$, c) $1 - (1 + t)e^{-t}$.

509. a) $\dfrac{24}{s^5} + \dfrac{2}{s}$, b) $\dfrac{1}{(s-1)^2} - \dfrac{1}{s}$, c) $\dfrac{3}{s^2+9} - \dfrac{s}{s^2+25}$, d) $\dfrac{1}{s-2} - \dfrac{2}{(s-2)^3}$.

510. a) $\dfrac{\omega}{(s+\alpha)^2 + \omega^2}$, b) $\dfrac{s+\alpha}{(s+\alpha)^2 + \omega^2}$.

511. $1/s$, e^{-Ts}/s, $(1 - e^{-Ts})/s$.

513. a) $(1 + s^2)^{-1} \coth \dfrac{\pi s}{2}$, b) $(1 + s^2)^{-1}(1 - e^{-\pi s})^{-1}$.

514. $\sqrt{\pi/s}$.

515. a) $\frac{4}{13} + e^{-2t}(-\frac{4}{13}\cos 3t + \frac{5}{39}\sin 3t)$,
b) $\frac{2}{85}\cos t - \frac{9}{85}\sin t + e^{-t}(-\frac{2}{85}\cos 3t + \frac{92}{255}\sin 3t)$,
c) $\frac{1}{2}e^{at}(\cosh t - \cos t)$, d) $\cos t + 1$, e) $(-1)^n t^n(a \sin t + b \cos t)$.

516. Zero.

518. Yes, the assumption that $f(t)$ belongs to the class E; the other assumptions imply that $f(t)$ has a Laplace transform.

520. a) $y = c_1 e^{t(1 + 1/\sqrt{2})} + c_2 e^{t(1 - 1/\sqrt{2})}$.
b) $y = c_1 e^t + c_2 e^{2t} + c_3 t e^{2t}$.
c) $y = c_1 e^{5t}\cos 6t + c_2 e^{5t}\sin 6t$.
d) $y = c_1 + c_2 t + c_3 e^t \cos t + c_4 e^t \sin t$.
e) $y = c_1 \cos t + c_2 \sin t + (t^2 \sin t + t \cos t)/4$.
f) $y = c_1 e^{3t} + c_2 e^{-t} - (24 \cos 2t + 23 \sin 2t)/65$.
g) $y = c_1 + c_2 e^{-2t}\cos t + c_3 e^{-2t}\sin t + t^2/10 - 4t/25 - te^{-t}/2$.
h) $y = c_1 e^{-t} + c_2 t e^{-t} - e^{-t}\log t$.
i) $y = c_1 \cos 2t + c_2 \sin 2t + \frac{1}{4}t \sin 2t + \frac{1}{4}(\cos 2t)(\log \cos 2t)$.
j) $y = c_1 \cos t + c_2 \sin t - 2 + (\cos t)(\log \cot t/2)$.
k) $y = c_1 e^{3t} + c_2 e^{2t} + e^{3t}\int_0^t e^{-3\tau}\tau \log \tau \, d\tau - e^{2t}\int_0^t e^{-2\tau}\tau \log \tau \, d\tau$.

521. a) $y = 2e^{-t}\cos t + e^{-t}\sin t$.
b) $y = \cos t - \sin 2t$.
c) $y = y_0 e^t + (y_1 - y_0)te^t$.

CHAPTER 6

602. a) $(2n - 1)!! 2^{-n} \sqrt{\pi}$, b) $(-2)^n [2n - 1)!!]^{-1} \sqrt{\pi}$, c) 0.

603. a) $\Gamma(x)/a^x$, b) $\Gamma(1 + 1/a)$, c) $\Gamma(a)$.

605. $y = c_1 J_{3/2}(x) + c_2 J_{-3/2}(x)$.

611. $y = x^{-1} J_0(mx)$ (or $y = c_1 x^{-1} J_0(mx)$).

612. $y = x^{-2} J_2(x)$ (or $y = c_1 x^{-2} J_2(x)$).

613. $y = J_0(x^m)$ (or $y = c_1 J_0(x^m)$).

614. $J'_{1/2}(x) = J_{-1/2}(x) - \dfrac{1}{2x} J_{1/2}(x)$.

616. $(1 + s^2)^{-1/2}$.

618. a) 0.353, b) -0.064, c) 0.224.

619. $-\sqrt{\dfrac{2}{\pi x}} \left(\sin x + \dfrac{\cos x}{x} \right)$.

620. $\sqrt{\dfrac{2}{\pi x}} \left[(3x^{-2} - 1) \sin x - \dfrac{3 \cos x}{x} \right]$.

626. $r_{1/2} = r_{-1/2} = r_{3/2} = \sqrt{\dfrac{2}{\pi}}$ and e.g. $\theta_{1/2} = 0$, $\theta_{-1/2} = \dfrac{\pi}{2}$, $\theta_{3/2} = -\dfrac{\pi}{2}$.

629. $x^p \sim 2 \sum_{n=1}^{\infty} [j_n J_{p+1}(j_n)]^{-1} J_p(j_n x)$, $0 < x < 1$.

630. $x^2 \sim 2 \sum_{n=1}^{\infty} (j_n^{-1} - 4j_n^{-3}) J_1^{-1}(j_n) J_0(j_n x)$, $0 < x < 1$.

636. $y = c_1 J_3(x) + c_2 Y_3(x)$.

637. (611) $y = c_1 x^{-1} J_0(mx) + c_2 x^{-1} Y_0(mx)$.

(612) $y = c_1 x^{-2} J_2(x) + c_2 x^{-2} Y_2(x)$.

(613) $y = c_1 J_0(x^m) + c_2 Y_0(x^m)$.

638. a) $y(x) = 3 J_2(x)/J_2(2)$,

b) there is no such solution.

641. a) $\dfrac{d}{dx} \left(x \dfrac{dy}{dx} \right) - \dfrac{p^2}{x} y + \lambda x y = 0$, $x \in (0, 1]$, where $p \geqslant 0$ is a given number,

b) $\lim\limits_{x \to 0} y(x)$ exists; $y(1) = 0$,

c) $\lambda_n = j_n^2$, $n \in Z^+$, where j_n is the nth positive zero of $J_p(x)$,

d) $J_p(j_n x)$, $n \in Z^+$.

CHAPTER 7

701. Let A denote an arbitrary function:

a) $z = \displaystyle\int_a^x f(t)\, dt + A(y)$, where $a \in D_f$,

b) $z = x f(y) + A(y)$.

702. $z = \begin{cases} A_1(x) & \text{if } y > 0. \\ A_2(x) & \text{otherwise in } \Omega, \end{cases}$

where A_1 and A_2 are arbitrary functions with the restriction that $A_1(x) = A_2(x)$ for $x < 0$.

704. a) $z = \dfrac{\partial z}{\partial x} \dfrac{\partial z}{\partial y}$, b) $\left(\dfrac{\partial z}{\partial y} \right)^2 = x \dfrac{\partial z}{\partial x} + y \dfrac{\partial z}{\partial y}$,

c) $z^2 - z \left(x \dfrac{\partial z}{\partial x} + y \dfrac{\partial z}{\partial y} \right) = 1$.

705. a) $x \dfrac{\partial z}{\partial y} - y \dfrac{\partial z}{\partial x} = x^2 - y^2$, b) $x \dfrac{\partial z}{\partial x} - y \dfrac{\partial z}{\partial y} = x - y$,

    c) $x \dfrac{\partial z}{\partial x} - y \dfrac{\partial z}{\partial y} = 0$,   d) $\dfrac{\partial z}{\partial x} + \dfrac{\partial z}{\partial y} = 0$.

706. $\left(\dfrac{\partial z}{\partial x}\right)^2 + \left(\dfrac{\partial z}{\partial y}\right)^2 = 4\left(x \dfrac{\partial z}{\partial x} + y \dfrac{\partial z}{\partial y} - z\right).$

707. $b \dfrac{\partial z}{\partial x} - a \dfrac{\partial z}{\partial y} = 0$. (If $P_0 = (x_0, y_0, z_0)$ is a point on a solution surface $z = u(x, y)$, then the normal vector $(u'_x(x_0, y_0), u'_y(x_0, y_0), -1)$ at P_0 and the vector $(b, -a, 0)$ are perpendicular.)

708. $x \dfrac{\partial z}{\partial x} + y \dfrac{\partial z}{\partial y} - z = 0$, $(x, y) \in \Omega$. I (Replace the vector $(b, -a, 0)$ in the previous answer by (x_0, y_0, z_0).)

709. $x \dfrac{\partial z}{\partial x} + y \dfrac{\partial z}{\partial y} = 0$, $(x, y) \in \Omega$. (Replace the vector (x_0, y_0, z_0) in the previous answer by $(x_0, y_0, 0)$.)

710. $A\left(\dfrac{y}{x}\right)$.

711. $A(\log x + y^{-1})$.

712. $z = x^{3/2} y$.

713. $A_1(2y - x^2)$ for $x \geq 0$ and $A_2(2y - x^2)$ for $x \leq 0$, where $A_1(u)$ and $A_2(u)$ are two arbitrary functions with the restriction that $A_1(u) = A_2(u)$ for $u \geq 0$.

714. Yes; $z = A(x^2 - 2y^2)$, where $A(u)$ is an arbitrary function with the restriction that $A(u) = -u$ for $u \leq 0$.

715. $A\left(\dfrac{x}{y}\right) - \dfrac{\cos xy}{2}$.

716. $A(3x + 2y) + x^2 + y^3$.

719. $A(e^{-x}(x + y + 1)) e^{x^2/2}$.

720. $ax + by$, where a and b are parameters.

721. $A(2x + y)e^{-3x} + x/3 - \frac{1}{9}$.

722. $z = (1 + \sin (2x + y))e^{-3x}$.

723. $A(x^2 + y^2)e^{c/y}$.

CHAPTER 8

801. The wave equation (1) of Problem 8.1.1 is obtained.

802. a) Yes, b) a unit of heat placed at the origin at time $t = 0$ dissipates in a rod of infinite length, c) no, for $u(0, t) \to \infty$ as $t \to 0$.

803. The Laplace equation (3) of Problem 8.1.3 is obtained.

804. The Laplace equation $u''_{xx} + u''_{yy} + u''_{zz} = 0$ is obtained.

806. a) Elliptic, b) parabolic, c) hyperbolic.

807. a) Elliptic for $4x > y^2$, parabolic for $4x = y^2$, hyperbolic for $4x < y^2$;
    b) parabolic for $y = 0$, hyperbolic for $y \neq 0$;
    c) elliptic for $xy > 1$, parabolic for $xy = 1$, hyperbolic for $xy < 1$;
    d) elliptic for $x^2 - y^2 > 1$, parabolic for $x^2 - y^2 = 1$, hyperbolic for $x^2 - y^2 < 1$.

808. $xu''_{xy} - yu''_{yy} - u'_y + 9y^2 = 0.$

809. For example: a) $\lambda_1 = 6, \lambda_2 = 1$, b) $\lambda_1 = 5, \lambda_2 = 0$, c) $\lambda_1 = 2, \lambda_2 = -3$.

810. a) Parabolic, b) neither (ultrahyperbolic).

811. $u(x, y) = (x^2 y - 2xy^2 + 2x + 2 \cos y)/(2 + 2x).$

812. b) No, $v(x, t) \to \infty.$

813. a) Yes, b) no (cf. Exercise 703), c) no. (The solution is not stable in the sense that a small change in the boundary condition produces an unbounded change in the solution.)

817. $u(x, t) = \frac{1}{2}[1 + (x + ct)^2]^{-1} + \frac{1}{2}[1 + (x - ct)^2]^{-1} + \frac{1}{c} \sin x \sin ct.$

818. $u(x, t) = \frac{8}{\pi} \sum_{n=0}^{\infty} (2n + 1)^{-3} \sin (2n + 1)x \cos (2n + 1)t.$

819. $u(x, t) = \frac{32}{\pi} \sum_{n=0}^{\infty} (2n + 1)^{-3} \sin (2n + 1)\frac{x}{2} \cos (2n + 1)\frac{t}{2}.$

820. a) $-11/192$, b) $-11/384$.

821. $u(x, t) = -8\pi^{-3} \sum_{n=0}^{\infty}(2n + 1)^{-3} \sin (2n + 1) \pi t \cos (2n + 1) \pi x.$

822. $E_k(t) = \frac{\pi}{2} \sin^2 t, E_p(t) = \frac{\pi}{2} \cos^2 t, E(t) = \frac{\pi}{2}.$

823. a) Parabolic, b) $u(x, t) = (x - t)f(x + t) + g(x + t), f \in C^2(R), g \in C^2(R),$
c) $u(x, t) = 2x^2 + 2xt$, d) yes.

825. $\frac{3}{64} + 1/(8\pi^2).$

826. E.g. $X''(x) + \mu X(x) = 0, Y''(y) + (\lambda - \mu)Y(y) = 0, T''(t) + c^2 \lambda T(t) = 0.$

829. $u(x, t) = \frac{\pi}{2} e^{-t} \sin x - \frac{16}{\pi} \sum_{n=1}^{\infty} n(4n^2 - 1)^{-2} e^{-4n^2 t} \sin 2nx.$

830. $u(x, t) = \frac{1}{2} - \frac{1}{2} e^{-4t} \cos 2x.$

833. $u(x, t) = (1 + 4t)^{-1/2} e^{-x^2/(1 + 4t)};$ yes.

835. $u(x, t) = 1 - 2\pi^{-1/2} \int_{0}^{x/\sqrt{4t}} e^{-z^2} dz;$ no, the function $v(x, t)$ of Example 8.2.4 can be added to $u(x, t).$

837. $u(x, t) = \frac{2}{\pi} \int_{0}^{\infty} \frac{\sin y}{y} e^{-y^2 t} \cos xy \, dy.$

840. $v(x, y, t) = \sum_{v=1}^{n} c_v J_0(j_v \sqrt{x^2 + y^2})e^{-j_v^2 t};$
$|v(x, y, t) - u(x, y, t)| < \varepsilon.$

841. $u(x, y) = \frac{4}{\pi} \sum_{n=1}^{\infty} (-1)^{n-1}(2n - 1)^{-2} \frac{\sinh (2n - 1)y}{\sinh (2n - 1)} \sin (2n - 1)x.$

842. $u(r, \theta) = 2\sum_{n=1}^{\infty}(-1)^{n-1}n^{-1}r^{-n} \sin n\theta.$

843. $u(r, \theta) = \frac{8}{\pi} \sum_{n=1}^{\infty} (2n - 1)^{-3}r^{2n-1} \sin (2n - 1)\theta.$

848. $u(r, \phi) = \frac{8}{5}r^3 P_3(\cos \phi) - \frac{3}{5}rP_1 (\cos \phi)$, where P_3 and P_1 are Legendre polynomials.

849. $u(x, y) = \sum_{v=1}^{4} u_v(x, y)$, where $u_1(x, y) = \sum_{n=1}^{\infty} c_n \frac{\sinh ny}{\sinh n} \sin n\pi x,$

$$c_n = \frac{2}{\sinh n} \int_0^1 f_1(x) \sin n\pi x \, dx, \text{ and analogously for } u_2, u_3, u_4.$$

850. $u(x, y) = \dfrac{4}{\pi} \displaystyle\int_0^\infty t(1 + t^2)^{-2} \dfrac{\sinh ty}{\sinh t} \sin tx \, dt.$

851. a) Cf. the introductory comment to Problem 8.4.6.

b) $z = (1 - \zeta)/(1 + \zeta)$, where $z = x + iy$ and $\zeta = \xi + i\eta$.

c) $u(r, \theta) = \dfrac{1}{2} + \dfrac{2}{\pi} \displaystyle\sum_{n=1}^\infty r^{2n-1}(-1)^{n+1} \dfrac{\cos(2n - 1)\theta}{2n - 1}$

$$u(r, \theta) = \frac{1 - r^2}{2\pi} \int_{-\pi/2}^{\pi/2} \frac{dt}{1 - 2r\cos(\theta - t) + r^2}.$$

d) $\xi^2 + \eta^2 = 1, \quad \xi > 0.$

852. No.

BIBLIOGRAPHY

1. L. V. AHLFORS, *Complex Analysis*, 2nd ed., McGraw-Hill, New York (1966).
2. A. ERDÉLYI and collaborators, *Higher Transcendental Functions*, Vols. I–III, McGraw-Hill, New York (1953–1955).
3. A. FRIEDMAN, *Partial Differential Equations of Parabolic Type*, Prentice-Hall, Englewood Cliffs (1964).
4. R. R. GOLDBERG, *Fourier Transforms*, University Press, Cambridge (1962).
5. K. HOFFMAN and R. KUNZE, *Linear Algebra*, Prentice-Hall, Englewood Cliffs (1961).
6. E. JAHNKE, F. EMDE and F. LÖSCH, *Tafeln Höherer Funktionen (Tables of Higher Functions)*, 7th ed., B. G. Teubner, Stuttgart (1966).
7. S. KACZMARZ and H. STEINHAUS, *Theorie der Orthogonalreihen*, Chelsea, New York (1951) (Warsaw 1935).
8. E. KAMKE, *Differentialgleichungen. Lösungsmethoden und Lösungen.* Vols. I–II, Akademische Verlagsgesellschaft, Leipzig (1959).
9. D. L. KREIDER, R. G. KULLER, D. R. OSTBERG and F. W. PERKINS, *An Introduction to Linear Analysis*, Addison-Wesley, Reading (1966).
10. W. R. LEPAGE, *Complex Variables and the Laplace Transform for Engineers*, McGraw-Hill, New York (1961).
11. I. G. PETROVSKI, *Ordinary Differential Equations*, Prentice-Hall, Englewood Cliffs (1966).
12. I. G. PETROVSKI, *Partial Differential Equations*, Iliffe Books, London (1967).
13. W. RUDIN, *Principles of Mathematical Analysis*, 2nd ed., McGraw-Hill, New York (1964).
14. G. SANSONE, *Orthogonal Functions*, Interscience, New York (1959).
15. L. SCHWARTZ, *Mathematics for the Physical Sciences*, Addison-Wesley, Reading (1966).
16. E. C. TITCHMARSH, *Eigenfunction Expansions, Part I*, Clarendon Press, Oxford (1962).
17. G. N. WATSON, *Theory of Bessel Functions*, University Press, Cambridge (1944).
18. K. YOSIDA, *Lectures on Differential and Integral Equations*, Interscience, New York (1960).
19. A. ZYGMUND, *Trigonometric Series*, Vols. I–II, University Press, Cambridge (1959).

References 1, 4, 5, 7, 9, 13, 14 and 16–19 have been cited in the text. References 2, 17 and 19 are very comprehensive; they probably treat everything essential that has been done in their respective fields up to the year of publication. In 8 there is a "dictionary" of solutions for some 300 first-order partial differential equations. In 9 there is a wealth of material and exercises on the subject of this book. The last chapter of 11 is devoted to first-order partial differential equations.

Anyone who desires a survey of the latest work in some branch of mathematics can consult the *Mathematical Reviews*, which since 1940 have been published by the American Mathematical Society in New York.

CONVENTIONS

We adhere to the following conventions unless the context clearly implies otherwise. Page numbers refer to the pages on which the conventions first appear.

The notation (a, b) is used when it is immaterial whether an interval under consideration is open, half-open, or closed, 1

In Chapter 1, functions are real-valued functions of a real variable (There are a few exceptions in Section 1.6.), 1

The notation $f(x)$ means a value or a function; the context shows the meaning, 2

Expressions such as "the function $y = \sin x$" and "the function $y = \sin x$, $0 < x < 2\pi$" are used to define functions, 2

Numbers within brackets refer to the bibliography, 2

To index finite sequences, Greek letters are used, 2

In Chapter 1, coefficients of a trigonometric series are real numbers, 5

Missing points on a curve are denoted by small open circles, 8

In Chapters 2–6, functions are complex-valued functions of a real variable, 26

The domain of a Laplace transform is not specified, 81

The domain of validity of an equation containing two or more Laplace transforms is not specified, 81

To denote a rational function and its inverse Laplace transform, $G(s)$ and $g(t)$ are used, 84

The values of an inverse Laplace transform $f(t)$ are not stated for $t \leqslant 0$, 85

An ordinary differential equation for which no interval is indicated, is given in the "largest" possible interval, 93

The symbols c_1 and c_2 denote complex parameters, 101

In Chapters 7 and 8, functions are real-valued functions of one or more real variables, 118

The letter u denotes a dependent variable or a function; similarly for z, 118

A partial differential equation for which no region is indicated, is given in the "largest" possible region, 120

A characteristic curve is as "large" as possible, 125

In boundary-value problems the boundary is approached along its normals; there are exceptions for the heat equation and the Laplace equation, 137, 148, 159

SYMBOLS

Page numbers refer to the pages on which the symbols first appear.

$\in$	belongs to, 1
R	the set of all real numbers, 1
$(a, b), [a, b], (a, b],$ $[a, b)$	(finite) interval, 1
$(a, \infty), [a, \infty),$ $(-\infty, a), (-\infty, a],$ $(-\infty, \infty)$	infinite interval, 1
f	function, 1
D_f	domain of f, 1
R_f	range of f, 1
$f(x)$	value of the function f at x; or the function f, 2
$\Rightarrow$	implies, 2
sup	least upper bound, supremum, 2
inf	greatest lower bound, infimum, 2
$\int_a^b f(x)\, dx$	definite integral from a to b, 3, 4, 26
f^+	positive part, 3
f^-	negative part, 3
Σ	sum notation, 5
$s_n(x)$	partial sum, 5, 29
Z^+	the set of all positive integers, 6
N	the set of all natural numbers, 6
Z	the set of all integral numbers, 6
$\sim$	has the Fourier series, 7, 29
$f(a^-), f(a^+)$	left limit, right limit, 14
D	class of functions, 14
Re, Im	real part, imaginary part, 23
$\lVert f \rVert$	norm, 28, 42
(f, g)	scalar product, 28, 42
I	index set, 28
$\bar{z}$	complex conjugate, 28
R^n	the space of all n-tuples of real numbers, 28, 118, 137
pr.v.	principal value, 29
$f'(x), f''(x)$	derivative, second derivative, 39

INDEX

ARNE BROMAN is professor emeritus of mathematics at Chalmers University of Technology, Göteborg, Sweden. He has had teaching positions from 1938 to 1981. He received his Ph.D. from Uppsala University in 1947 for a thesis on trigonometric series. He was research associate at the University of Chicago in 1948. He has published research papers, mostly on classical analysis, and several high-school and college-level texts in Sweden. In 1952 he published, in a Swedish science series (the Verdandi series), a monograph on famous mathematical problems. In the late 60's and early 70's he taught for three years at Simon Fraser University and Western Washington University. In 1976 he produced a film on four-dimensional geometry, and in 1980 he was a member of an expedition from Chalmers to Peru to study Inca roads. In 1983 he got a Techn. D. at Chalmers for a thesis on road alignment. In addition, he has published several papers on solar-energy collectors in cooperation with his eldest son.